Cosmic Force Cosmology
21st Century Scientific-Philosophic Revolution Manifesto

By Solatle Lu

SECOND EDITION

Published by New Generation Publishing in 2022

Copyright © Solatle Lu 2022

Second Edition

The author asserts the moral right under the Copyright, Designs and Patents Act 1988 to be identified as the author of this work.

All Rights reserved. No part of this publication may be reproduced, stored in a retrieval system or transmitted, in any form or by any means without the prior consent of the author, nor be otherwise circulated in any form of binding or cover other than that which it is published and without a similar condition being imposed on the subsequent purchaser.

ISBN
        Paperback   978-1-80369-3514
        Hardback   978-1-80369-3521

www.newgeneration-publishing.com

Contents

My Dream 1
Preface of the Second Edition 3
Preface of the First Edition 7
Anthem of Vortex Motion 13
The Universe Has No Secrets 17
Is the Essence of the Universe Material or Energetic? 20
The Cosmic Force Is the Aggregate Force of Particles ... 22
The Cosmic Spirit is an Absolute and Infinite Creativity 24
Darkened Light Theory (DLT) 28
Light Ball Theory (LBT) 32
Light Field Theory (LFT) ➤ Light Ether 34
Two Scientific Research Roads: Holism vs Corpuscularianism 39
Three Types of Corpuscular Theory 46
The Quintessence of Holism: Light Field Theory & Vortex Theory 51
All Fields Transmit Energy in Waves 55
Enhanced Energy Effect of Field (EEEF) 58
$1\oplus + 1\ominus = 30{,}000$ 61
Important Scientific Discoveries Results from Particle Colliders 64
Antimatter without a Foothold 69
Ubiquitous Antiparticles 71
New Interpretation of the Wave–Particle Duality of Light 74
All Particles Are Derived from Photons and Live Forever 77
The Speed of Light is the Boundary Between Matter and Energy 83
Mass-Energy Conversion: Changing Speed with Mass ... 86
Energy-Mass Conversion: Changing Mass with Speed ... 90
Cracking the Myth of the Double-Slit Experiment 93
History of the Vortex Theory 96
Absolute Truth: Cosmic Force Vortex Theory (CFVT) 102
The Direction of the Hubble Tuning Fork Is Wrong 105

Constant Speed of Light ➤ Doppler Effect Becomes Ineffective 107
  Light Waves Get Longer as They Travel and All Starlight Is Redshifting 110
  The Differences Between the Ground Vortex and the Sky Vortex 114
  The Galactic Density Wave Theory Lacks Persuasion 116
  Spiral Shell Theory (SST) 118
  The Role of the Shell in the Formation of Stars 120
  The Orbital Velocities of Planets Show the Vortex Effect 122
Shells Are a Stable and Orderly Factor in the Cosmos .. 126
Why Can't Planets Run Away? 128
Why Vortex Force and Shells Frequently Cause Explosions 130
Three Vortex Theories: Descartes', Newton's and CFVT 132
Orion: The Baby Star Garden 144
  A Sky of Fire and Ice 147
  Firelight: Fire Is Different from Light 150
  The Sun Shines but There's No Fire—No Nuclear Fusion 154
  The Formation of the Sun and the Solar System 156
The Mechanism that Forces a Comet to Turn Around .. 162
The Gravity of the Sun Is Weaker Than That of the Planets 164
  How Strong Is the Earth Gravity? 166
  Gravity and Tension 168
  New Definition of Gravity 171
  The Forces to Fix Things on the Earth Surface 174
  Factors that Constitute the Weight of Ground Objects 176
  Jupiter Is a Small Star That Glows Intermittently 179
  The Misunderstood Group: Jovian Stars 182
  The Causes of Saturn's Low Density 185
  Tides Come from the Sloshing Effect of Seawater 187
  A New Astronomical Law Arising from the Slow Spin of the Moon 192
  Comparison of Rotational Motion and Vortex Motion 196

Methods to Judge the Spin Motion Properties of Stars 198
Evolution of the Earth 201
Solar System Q & A 203
The Most Common Products of Vortex Motion: Spheroids
.... 210
Why Scientists Ignore the Greatest Forces in the Universe
.... 212
  Query One: Why Can't We See the Winding? 214
  Query Two: Why Do We Not See the Shortening and
  Disappearance of Spiral Arms? 216
  Query Three: A Lot of Stars Have Been Observed
  Moving Faster Than the Spiral Arms 219
  A New Theory of Black Hole Genesis and Evolution 221
  How Vortexes Make Black Holes Easily 223
The Black Hole: The Smelting Furnace of Celestial Bodies
.... 226
  Quasars Are the Outer Shells of Spiral Black Holes in
  Galaxies 228
  Black Hole Pulses Were Mistaken for Neutron Stars . 231
A Supernova Explosion Is a Phenomenon of Galaxy
Renewal 235
  Vacuum Is a Virtual Space, Physical Space Is Warped
  237
  Design a Dynamic Star Diagram 241
  Particle Physics: Fireworks Theory 243
  Two Entities of the Universe: Energy and Matter 246
  The Principle of Division and Union 249
  The Atomic Bomb Is Also a Fusion Weapon 252
  Cosmic Energy and Material Energy 254
  Cosmic Field and Material Fields 257
  GUT of Energy Achieved by Particle Vibration 263
  Radiation Is a Universal Characteristic of Objects 268
The Mechanism of the Self-Preservation of Everything:
Shells 273
  The Main Force of Antigravity Is Radiation 275
  Magnetism and Gravity Disappear at High Temperatures
  278
  New Mode of Gravity: Tug-of-War 280

A Tug-of-War between the Earth and the Moon......... 283
Gyro Effect, the Answer to the Riddle of Celestial Body Revolution and Spin... 286
What Would Happen if the Sun Disappeared? 290
Astronomical Facts Opposing the Big Bang-Cosmic Inflation Theory 291
   A New Look at the Big Bang... 294
   The Crisis of Quantum Physics..................................... 297
   The Essence and Prospect of Quantum Theory 302
Quantum Theory May as Well Be Replaced by Photon Theory 305
   No Messenger Particles... 307
   The Duality and Three Conditions of Things 311
   Matter Is Atom 316
   Matter Is Trapped Particles.. 318
   Two Kinds of Bodies: Ionic Bodies and Atomic Bodies 323
   Sunspots Are Windows to the Sun Hypothermic Spheres 325
   How Can Plasma Be Preserved in Stars permanently?330
   Plasma ➤ White Dwarf ➤ Black Dwarf ➤Nuclear Bomb.. 332
A Nuclear Bomb— a Miniature Sun Ready to Explode 335
Genesis Conditions 341
The Way of Mass-to-Energy Conversion 344
Atoms, Corpuscles, and Reductionism........................... 346
Utilization and Conditions of Material Energy.............. 349
Spontaneous Human Combustion (SHC) — Ions & Quarks Revel in Freedom... 352
Dark Matter Is Mainly Composed of Nebulae, Small Stars, and Black Holes... 359
In the Simplest Sense, Dark Energy Is Light and Black Holes 361
Black Holes Are Mass-Energy and Energy- Mass Converters... 365
Nothing but Living Things Really Die 367
The Goal of the Grand Unification: The Unification of the Speed of Forces 370

Scientists Have Been Exploring Electricity and Magnetism for 400 Years 375
Electricity, A Common Widespread Misconception 378
The Simplest and Clearest Principles of Electricity & Magnetism 387
  Magnetic Field-Electric Field-Ion Field — the Triad of Electricity 392
  LFT vs Relativity 397
Two Ends of the Grand Unification: Holism vs Corpuscularianism 404
Solutions: Newton's Five Big Puzzles 407
Paraphrase Newton's Forces 415
The Road to Ultimate Theory 419
Wisdom Is the Author of Truth and Knowledge 424
The Task of Science: Discovery or Design? 427
Big Bang Model vs Cosmic Force Model 429
A Strong Pillar of CFT ➤ Systems Science.................. 437
The First Step of Creation: Choosing a Direction 443
Cosmic Evolution and Human Mission 446
Why Do I want to Publish This Book? 449
Appendix: Firmly Support Absolute Justice ACT (AJA), Strongly CurbNuclear War! 456
References 460

My Dream

As the world entered the 21st century,
Aged Solatle Lu one day had a young dream:

Give me downright facts and I will create a cosmos out of them!

Three years later,
I got innumerable downright facts from astronomers and webs,
My dream came true.
The Cosmic Force Cosmology was born
— An absolute and eternal truth;
The intelligent light brighter than the brightest starlight in the sky.

Preface of the Second Edition

This book is an enhanced version of *An Outline of Force Cosmology* (2019) & *Cosmic Force Cosmology—21st Century Scientific-Philosophic Revolution Manifesto*. (2021)

Cosmology has always been the category of reductionism, and this book can be deemed to **the first deductive cosmology**. It draws a new picture of the cosmos, and is a science as well as a philosophy work. In science I dubbed it cosmic force cosmology (CFC), and in philosophy, cosmic force theory (CFT). In terms of academic division, cosmology is philosophy, whereas astronomy, which is akin to cosmology, science. They can be paraphrased as theoretical astronomy and technical cosmology respectively. **Their relations are fixed. Cosmology is director, astronomy, assistant. Astronomy can be real only under the guidance of real cosmology.** The history of world culture shows that only deductive and speculative philosophy can produce real cosmology.

Astronomy is one of the 'seven liberal arts' in ancient Greece, and later world naturally pay more attention to astronomy than cosmology. This book is a synthesis of the two disciplines.

Although famous theories have been linked to cosmology, hence Ptolemy's cosmology, Copernicus's cosmology, Newton's cosmology, Hubble's cosmology, Einstein's cosmology, etc., and standard cosmology (SC) has come out of them, yet there is no any book with 'cosmology' title in the recommended books catalogue on the Internet. Maybe someone adopted this title but didn't attract attention. Anyway, the title of this book is unusual, not just as a unique cosmology, but as a manifesto for a revolution.

Cosmos is the highest and grandest being known to man,

and cosmology is rightly the highest knowledge, the knowledge of everyone, especially in today's integrated world.

Looking back at the history of civilization, we find that even the fore-mentioned fragmentary cosmologies have had a huge and profound impact on the world, not limiting to scientific research, but involving group behaviors and our world views.

The world has always been governed by the reductionist world views, which sees everything as separate particles, what awaits this world is the entropy maximum and disasters. People live in confusion and anxiety, waiting for the arrival of the end of the world. "The mechanistic world view, taking the play of physical particles as ultimate reality, found its expression in a civilization which glorifies physical technology that has led eventually to the catastrophes of our time." (Bertalanffy, 1969, p49)

CFC is an antithetical cosmology. It paints a picture of a vibrant cosmos. Nearly 400 years ago, Galileo Galilei, the central figure of the scientific revolution of the seventeenth century, started *Dialogues Concerning Two New Sciences*. This book is a Dialogue between SC & CFC.

SC
- The cosmos is material. The singularity explosion produces 10^{80} particles and a much larger number of messenger particles that make up the universe, whose mass and energy are provided by the God particles.
- **Matter is infinitely divisible**, a series of molecules, atoms, subatoms, strings...
- Correspondingly, the space is also a series: one-dimensional, two-dimensional, three-dimensional, four-dimensional... **eleven-dimensional.**
- A new discovery is **dark matter**, a strong gravitational pull. It is several times more than matter, and its nature is unknown.
- Another new discovery is **dark energy**, a powerful

tension force, which accounts for more than 90% of all matter & energy, and its nature is unknown.
- Four fundamental forces dominate the motion of all things. The strong force, weak force and electromagnetic force are short-range forces, mainly acting between particles. **Gravity is the weakest but strongest long-range force, dominating all celestial motions.** Their forces come from gluons, photons, virtual photons and other messenger particles, and gravitons yet to be discovered.

CFC.
- The cosmos is **LIGHT,** which is the continuum of energy & mass. Free light is energy, an infinite & eternal cosmic energy, its main form is a light field; Imprisoned light is matter, relying on external forces to move. Both free light & imprisoned light are unified in vibrations, high & low frequencies.
- Matter is not infinitely divisible; **its limit is the speed of light.** When moving at the speed of light, it is no longer matter but energy.
- **Matter-energy reciprocity** is the most basic and universal movement in the universe.
- Space is three dimensions plus time is **four dimensions.** There is no higher dimensional space. **Each object has its own time-space**, which overlaps each other, and the time-space of the high-level system embraces that of its subjects. The time arrow always forward, irreversible; Each space is non-transferable and never reproduces. **Cosmos energy (ether) is independent of time-space, it is the eternal cosmic background**.
- **The most pervasive and powerful force in the universe is the vortex motion that generates everything,** including dark matter & dark energy, and most notably, plasma. All the elements, stuffs, and living things on earth evolve from plasma.

The highlight of CFC is the distinct humanism, to elevate the status of human beings in the cosmos to an unprecedented height. It claims that **Man is the soul of the cosmos,** emitting more brilliant light of intelligent than starlight. Man is a spectator as well as a performer, his fundamental mission is to create things that Nature can't, to turn our planet into an ideal world, and to push the evolution of the cosmos to its highest stage.

Man has many weaknesses and defects, even cruel and wick. But there is something sublime and holy about man. "What a piece of work is a man! how noble in reason! how infinite in faculty! in form and moving how express and admirable! in action how like an angel! in apprehension how like a god! the beauty of the world! the paragon of animals!" (Shakespeare, *Hamlet*) "We are the miracle of miracles, the great inscrutable mystery of God." (Thomas Carlyle) In material cosmos, man is insignificant; And in spiritual cosmos, man's intelligent light overwhelms the brightest starlight. It is man who elevates the universe from an irrational being into a rational being.

No doubt, through long and fierce struggles and continuous creative activities, **beauty and good will triumph over ugliness and evil.** And mankind will be able to survive far beyond the dinosaur did, reaching 200 million to 1 billion years, has plenty of time to accomplish his mission.

Solatle Lu, Kunming, June 2022

Preface of the First Edition

Generally, downright fact may be told in a plain way; and we want downright facts, at the present, more than anything else.
John Ruskin (GIGA Quotes)

Sir Isaac Newton valued phenomena and believed that any proposition that was not derived from them should be called a hypothesis, which is not advisable by experimental philosophy. This is also a faith commonly held by current scientists.

When a phenomenon is confirmed it becomes a fact and the universal fact, without exception, is truth. Nineteenth century German philosopher Friedrich Nietzsche said, "There are no eternal facts, as there are no absolute truths." (Goodreads Quotes) If it's true, then a theory founded on eternal facts or downright facts is Absolute Truth. Cosmic force vortex theory (CFVT) is a theory that is based on eternal facts. "Deductive reasoning, then, consists of those ways of deriving new statements from accepted facts that compel the acceptance of the derived statements... Deduction, as a method of obtaining conclusions, has many advantages over trial and error or reasoning by induction and analogy. The outstanding advantage is the one we have already mentioned, namely, that the conclusions are unquestionable if the premises are... With all of its advantages, deductive reasoning does not supersede experience, induction, or reasoning by analogy. It is true that 100 per cent certainty can be attached to the conclusions of deduction when the premises can be vouched for 100 per cent." (Kline, 1953, p27)

Astronomers tell us that **the sky is full of vortexes.** Among them, the most conspicuous are countless spiral galaxies, without exception, which therefore makes this an eternal fact. Common sense tells us that **all vortex motions are coaxial, with no exception,** and this is another eternal fact. CFVT established in these two facts may be called the Absolute Truth — of all existing theories, it is the sole theory supported by infinite eternal and downright facts.

That's not all. Each of the spiral galaxies we observe dominate hundreds of billions of stars, sending them toward the centre of the galaxy with spiral arms that eventually twist them into the central black hole. From this phenomenon comes another eternal and downright fact:

vortex force is the greatest force in the universe. Strong force, weak force, electromagnetic force and gravity, are all insignificant compared with vortex force, and their efficacy is weak or lost in it. Physics, chemistry and astronomy have to be rewritten, and our worldview has to be renovated — that is the implication of the subtitle of the book.

We can't deny reductionism completely, it leads people to go deep into things and see their structure, essence clearly. Its greatest achievement is to help scientists discover atomic energy and the way to release this energy.

However, we cannot ignore the defects of reductionism, which pays more attention to debris and neglects the whole, and thus tends to lose the right direction.

The reductionist method, for example, is to ask people to smash a mirror and grind it into powder, then use the world's largest machines (particle accelerator & particle collider) to observe and analyze these particles and their shadows (quantum), and find new particles and strange particles, to realize the unity of particles and achieve the truth of mirror.

People always adore and trust scientists, appreciate their efforts, reward their outstanding merits and make the best use of their research results.

However, when they build more and more machines, which are bigger and bigger, even too big for the Earth to accommodate, and empty the coffers of all countries on the world that can't support it (Hawking, 2016), I show them another way: **Holism. It's a simple and precise way,** and I hope that some of them can pay attention to it. This is also a Greek work, and Plato's theory of ideas is the best holism. Idea is the prototype and ideal state of things, that is, the essential and perfect state of things. Plato pointed out that our study should begin not only with the big and obvious, but also with the ideal. **When we recognize the ideal type of things, like beauty itself, we gain the vision of the soul, and we can see at a glance the imperfect and defective.**

Focusing on an ideal mirror is the simplest and most effective way to recognize a mirror. From this mirror, we can easily identify the imperfect mirror and the defective mirror. From it we see not only what we really look like, but also know some characteristics of glass, mercury and their particles that makes it up — All this knowledge you can't get from reduction, and you can't find the way to the truth of the mirror from its particles, not to mention the shadow of particles — quantum.

Idealism entails people to know not only the beauty of all things, but also beauty itself, and then the goodness. Understanding beauty and goodness is the goal of philosophy and cosmology. In philosophy, Plato has done pioneering and foundation work, but it is not perfect. I intend to continue his work, which is the purpose of my writing *The Plato Code*, and *Cosmic Force Cosmology* (CFC) is an innovation based on it.

Compared with SC, the most outstanding feature of CFC is its view of the cosmic force, which is infinite and unique: **LIGHT.** The root of all energy, matter and matter motion is light, the various forms of photons, and the universe is the evolution process of light.

Its argument comes from one of this book's dogma: **Matter is trapped particles, or quarks, or photons, or energies;**

energy is free particles, or free quarks, or free photons. Matter and energy are united in this way.

The smallest unit of matter is the atom. Protons in atoms are considered to be made up of three quarks, the most basic and smallest system in the physical world.

A few scientists may think that a quark is an energy particle, but most of them, most quantum scientists, think that a quark is a matter particle — "What happens is that a **matter particle**, such as an electron or a quark, emits a force-carrying particle." (Hawking, 2016, p.78). In particle physics, bosons are force-carrying particles, and fermions are particles constituting matter. Particle physicists believe that not only Proton and neutron but also electron & quark are fermions.

The uniqueness of cosmic force theory (CFT) lies here. It holds that **photons are a continuum of energy and matter**, determined by its velocity. When it travels at the speed of light, the free photon, zero mass, is energy; The mass obtained by trapped photons is matter. Quarks are a variety of photons. Free quarks are energy. While they are trapped and not allowed to move, they are matter. The smallest unit of matter is atom, inside it is energy, infinitesimal fields of energy, energy and matter are thus united. Therefore, the above dogma can be written as:

Energy is free quarks, and matter is trapped quarks, or trapped energy.

In *A Brief History of Time*, Stephen Hawking presented readers with a photograph and wrote: "Fig. 5.2 shows a photograph of a collision between a high-energy proton and antiproton, producing a couple of almost free quarks." (p.84) The figure provided by Hawking is evidence of mutual change of mass-energy, though I don't agree with the antiproton view. According to CFT, Proton & antiproton are matter; free quarks, energy.

From the above discussion, it is easy to understand the phenomenon of mass-energy interconversion, and it is inevitable that this reciprocal event is happening everywhere and all the time. It is the foundation of CFC in the microscopic world. And its foundation in the macroscopic world is the spiral galaxies and other less vivid celestial vortexes.

At this point, the difference between CFC and SC becomes obvious. The two are opposed not only to most ideas, but to systems: holism vs reductionism. The vortex motion and the energy connotation of atoms, which are the basis of CFC, are downright facts, and thus become an absolute truth. While in SC, there are not many definitive facts that we can find, even gravity and quantum are fuzzy, spooky facts.

Frederick Engels wrote in the Introduction of *Dialectics of Nature* about 140 years ago:

"Thus we have once again returned to the point of view of the great founders of Greek philosophy, the view that the whole of nature, from the smallest element to the greatest, from grains of sand to suns, from protista to men, has its existence in eternal coming into being and passing away, in ceaseless flux, in un-resting motion and change, only with the essential difference that what for the Greeks was a brilliant intuition, is in our case the result of strictly scientific research in accordance with experience, and hence also it emerges in a much more definite and clear form."

Here I would like to add that **vortex motion is also in ceaseless flux, in un-resting motion and change, but it is the solely constant and eternal form of motion in the universe.**

CFC has a higher goal, and that is humans. The ultimate goal of our study of the universe is to recognize man's position in the universe and man's mission to improve and elevate their species.

First, we need to make it clear that **man is the universe,** and the ultimate purpose of studying the universe is to know man. As Eckhart Tolle, the most inspiring and visionary spiritual teacher in the world today said, "You are not *in* the universe, you *are* the universe, an intrinsic part of it. Ultimately you are not a person, but a focal point where the universe is becoming conscious of itself. What an amazing miracle." (Nomadrs Quotes)

Secondly, **man is the most exquisite and intellectually advanced product of the universe – the soul of the universe.**

There is only motions, changes and creations in the universe, and no words. Man naturally becomes the spokesman and promoter of the universe. As the British 20th century philosopher Alan Watts said, "Through our eyes, the universe is perceiving itself. Through our ears, the universe is listening to its harmonies. We are the witnesses through which the universe becomes conscious of its glory, of its magnificence." (Goodreads Quotes)

Man is not merely witnessing the glory of the universe, he is adding to it, making it even more brilliant by creating things that were not in the universe before, including fine music, paintings, sculptures, industrial products, buildings, and CFC is the latest one.

F. Engels asserted that although the existence of human beings is short compared with celestial body motion, our efforts will not be in vain. Because the human mind is an integral part of the eternal cycle of the universe, the immortal soul of it.

"It is an eternal cycle in which matter moves, a cycle that certainly only completes its orbit in periods of time for which our terrestrial year is no adequate measure, a cycle in which the time of highest development, the time of organic life and still more that of the life of beings conscious of nature and of themselves, is just as narrowly restricted as the space in which life and self-consciousness come into operation... **that none of its attributes can ever be lost, and therefore, also, that with the same iron necessity that it will exterminate on the earth its highest creation, the thinking mind, it must somewhere else and at another time again produce it."** (*Dialectics of Nature*)

Solatle Lu, Kunming, Oct. 2021

Anthem of Vortex Motion

Explorers who have come so far, please stop and listen to a thinker. Your direction is turning—
in front of you is boundless fog and mud.
The further you go, the farther you get from Sunshine Avenue.

Look up to heaven.
The grand stars have long pointed out a broad way.
Pay attention to the magnificent celestial vortex structure everywhere.
Don't listen to such superficial talk as 'spiral arm pattern', Firmly believe that it is an exquisite work of nature over ten billion years—
the display of the greatest force of the universe.

You needn't move feet, just roam with your eyeballs, and continue to look down the spiral arm toward the centre of the vortex.
More glorious and magnificent scenes will appear.

The rotating arm is like a raw material conveyor belt of stars 'smelting furnace'.
Huge stars are rolling toward the centre, one by one.
At the tapering end of the spiral arm, they are torn into long, wide strips.
Their powerful energy is instantly released, dazzling, dazzling light for billions of miles.
They are quasars that have amazed and puzzled astronomers

Inside the shining shell is a giant black hole rotating at high speed.
It is the insatiable 'smelting furnace' body.
Quasars will stall and get involved with the furnace after a brief bit of glory,
making black holes grow stronger and stronger, heating up and pressurizing.
But its noumenon is always dark and invisible.

The black hole, created by the vortex, spins at the speed of light.
It turns the quasars, emitting strong high-frequency signals, which are mistaken by astronomers as pulsar or neutron star.
Quasar, pulsar, black hole are really a trinity.

Don't go away.
Keep your eyes on this ever-growing-larger black hole.
When it devours almost all bodies in the whirlpool,
the spiral arms disappear.
Under the endless and irresistible twisting force, continuous encryption, temperature rise, and pressurization, energy-mass conversion occurs.
The black hole is suddenly bright and brilliant. A resplendent superstar is born!

Ah, there is too much energy.
The self-luminosity and heat still cannot be dispelled.
Vortex force makes it continue to shrink, pressurize, and heat up.
Violent and continuous explosions are the only way out.
And so, you see, the supernova explodes and shines for hundreds of millions of miles.
Next may be the birth of binary stars, three stars, star clusters, and even a new galaxy.
Vortex movement is everywhere, and that's what you can see the galaxy vortex.
What's more invisible is the vortex of stars and the vortex of planets.
And many tiny swirls — satellites — are created by this swirling motion.

Looking up, you will find the bigger spiral of a super-galaxy. The spiral arm itself is also in the process of winding and contracting.

The universe is just a maelstrom; big vortexes set small vortexes in motion.

All products produced by the vortex motion are small and large round plasma.

Only after cooling down can solid spheres, such as Mercury, Venus, Earth, Mars, and the moons, be found.

Vortex motion also creates large and small energetic fields, mainly light fields.

It expands out into space at twice the speed of light, indefinitely. This is another universal and magnificent sight in the sky.

In front of the vortex force and light field,

gravity appears, weak and irrational — no one had yet thought that gravity would produce antigravity.

Where does the power of vortex motion come from?

It's neither God's first push, nor something out of nothing.

You may not believe

It's the energy of countless particles at the expense of velocity. The vortex drops gather together and make the most of them. A trillion bodies are pouring out.

And by the inviolable will of the light field,

the vortex force thus becomes the absolute and infinite material creation.

By understanding the principles of the vortex,
and by understanding the extraordinary theories of black light, light ball, and light field,
strange changes of every celestial body,
all the mysterious astronomical phenomena,
will become clear and reasonable and easy to understand.
The truth of the universe will be revealed to you.

Praise thee, eulogise thee,
that never ends and never fails vortex force—
the greatest and most powerful creativity in the universe, giving impetus to all things in the universe,
impelling the movement and evolution of everything!

The Universe Has No Secrets

In short, there is nothing in the whole of nature (nothing, that is, which should be referred to purely corporeal causes, i.e. those devoid of thought and mind) which is incapable of being deductively explained on the basis of these selfsame principles; and hence it is quite unnecessary to add any further principles to the list.
René Descartes

What blessedness it is to dwell amidst this transparent air, which the eye can pierce without limit, amidst these floods of pure, soft, cheering light, under this immeasurable arch of heaven, and in sight of these countless stars! An infinite universe is each moment opened to our view. And this universe is the sign and symbol of Infinite Power, Intelligence, Purity, Bliss, and Love.
William Ellery Channing (GIGA Quotes)

Although the universe is vast and boundless, it keeps no secrets from human beings. Because we live in it, the real shape and movement of the universe is always present and on display, which are strict, stable, and orderly, unchanging for hundreds of millions of years.

Another advantage is that we are based on a planet with the richest species and the most diverse environment. The most complex structure, the most perfect species, and the most magnificent scenes in the universe all appear on the Earth or can be easily observe from the earth. Man himself is the highest and most wonderful sort of all things. We embody the most incredible spirit of the universe.

In *Universal Natural History and Theory of the Heavens*, the eighteenth-century Prussian philosopher Immanuel Kant said that the structure of the universe is simple, not supernatural (p.31). Among the origins of various natural objects studied by men, the origin of the cosmic system, the origin of celestial bodies, and the causes of their motion are expected to be thoroughly and correctly understood first (p.16). Also, the causes of the formation and motion of all celestial bodies — the structure of the whole universe — can be recognized first. It is much easier than using mechanical reasons to completely and clearly explain the generation of a weed or a larva (p. 17).

However, scientific exploration is a free field with no constraints and has many participants. **Some people are used to seeing complex things as simple and the universe as an understandable entity, while others are accustomed to seeing the simple as the complex, making the universe mysterious and incomprehensible.** Both have many followers. The successor needs knowledge and, more importantly, wisdom to make the right choice between the two groups. Following Francis Bacon & Galileo Galilei, scientists have kept to the path of knowledge. They believe that knowledge is power and that experiments and mathematics are the bridge to truth. It is not until Albert Einstein that we hear a different voice: imagination is more important than knowledge. This book goes a step further, claiming that wisdom is our guide to truth.

For most people, experiments and mathematics are useful and difficult, but they are just tools of exploration. **We should not let these tools stifle our imagination and creativity.** Von Hohenheim Paracelsus, Michael Faraday, Thomas Edison are our role models. They lack mathematics background, and their works rarely use mathematics. They employ simple and rude equipment to carry out scientific experiments, but show the mysterious alchemy and electromagnetic theory in the most intuitive ways, benefiting mankind.

Einstein said: "Do not worry about your difficulties in Mathematics. I can assure you mine are still greater." If that doesn't reassure you, listen to Thomas Edison: "An essential of an inventor is a logical mind that sees analogies. No! No! not mathematical. **No man of a mathematical habit of**

mind ever invented anything that amounted to much. He hasn't the imagination to do it. He sticks too close to the rules, and to the things he is mathematically sure he knows, to create anything new." (As quoted in French Strother, 'The Modern Profession of Inventing')

René Descartes began his academic research by doubting the existing theories, and believed that the writer should be "always starting with the simplest and easiest things and never going beyond them till there seems to be nothing further which is worth achieving where they are concerned," (Descartes, 1985, p.20) so that non-professionals can have intuitive cognition at a glance. He criticized those mysterious practices. "They may have feared that their method, just because it was so easy and simple, would be depreciated if it were divulged; so to gain our admiration, they may have shown us, as the fruits of their method, **some barren truths proved by clever arguments,** instead of teaching us the method itself, which might have dispelled our admiration." (Ibid., p.19)

Harvard requires its students to be friends with Plato, Aristotle, and even more so with truth.

This book advises readers: **Be friends with truth, with knowledge, but most of all, with wisdom. Wisdom is the author of truth & science.**

Is the Essence of the Universe Material or Energetic?

Men do not know how what is at variance agrees with itself. It is an attunement of opposite tensions, like that of the bow and the lyre.

Heraclitus, *Fragments*

The universe has no mysteries and is comprehensible, but that does not mean that everyone can comprehend it easily. A fundamental problem is the nature of the universe. If we misjudge it, the established cosmology may be wrong. The current popular cosmology, that the universe is based on matter and gravity, is wrong. **Contemporary scientists have found that the proportion of dark energy in the universe is far greater than that of matter, which also contains tremendous energy inside. Energy is the essence of the universe, and matter must originate from it, not vice versa.**

CFC is based on this judgment. I call it an energy-based theory (EBT). The existing cosmology, a matter-dominated cosmology, I call it matter-based theory (MBT), and its representative work is big bang cosmology.

Energy has no specific form, which is invisible and difficult to measure. Most materials are concrete, tangible, visible and measurable, so MBT has been established and developed very early, whereas the EBT has never been established until this book.

The basic difference between the two theories is in the reciprocity of energy and mass. Although MBT recognizes the principle of reciprocity, it is considered to be very difficult and rare. Because everything is made of particles, they interact by exchanging messenger particles, and have no energy of their own; they cannot be converted to energy, especially mass particles. In addition, all the particles are the particles left over from the collision of positive-negative particles at the moment of the Big Bang, and the number is fixed. So far, the way to produce matter from energy has not been discovered.

Quantum theory claims that a light quantum is a photon and not a particle, and that it is a field with a wave. "The photon is a spread-out field; the particle-like behaviour occurs because each photon, or quantum of field, is absorbed as a unit." (Brooks, 2016, p.68) In this description, a quantum looks more like a particle shadow than energy. Quantum theory is the energy theory based on the MBT.

MBT supports the expansion of the universe, and holds that dark energy is pushing the expansion of the universe, while dark matter forms gravity and blocks the expansion of the universe. However, the big bang 13.7 billion years ago was unstoppable, and the universe has been expanding and accelerating all the time.

EBT holds that **particles are energy & mass unity, that the two mutual change occurs at all times.** Energy particles trapped in atoms aggregate into tiny fields of matter (proton), and energy particles half trapped in stars aggregate into huge fields of matter (plasma), all of which are strong fields. The outer shell of all things is matter, and inside is energy. Free particles outside of atoms, molecules and objects make up the cosmic background energy field, implicating the universe is a life process with endless vitality. EBT thus belongs to the optimist, while MBT, the pessimist.

The Cosmic Force Is the Aggregate Force of Particles

All particles have the essential characteristics of flight, spin, and vibration, which are manifestations of force, the innate & inherent force, not foreign. I call the aggregate force of particles the cosmic force, so is the impetus of everything. Force will only change from one form to another, never fade away.

Physics tells us that motion is an attribute of matter. Particles fly in space, electrons surround nuclei, planets surround the Sun, and the Sun surrounds the centre of the Milky Way. All of them are moving eternally. At the same time, they are constantly spinning, vibrating and each has its own orbit, showing regularity. Theses motions are beyond doubt.

However, physics fail to tell us where the driving force of all things comes from. Both strong and weak forces are short-range forces that act inside the object but have no effect on the outside of the object. Electromagnetic force and gravitation are considered as long-range forces, but they are linear forces and cannot be applied to the curvilinear motion and orbital motion of all things. In addition, these two forces decay rapidly as distance increases, and gravity is the weakest of the four. In desperation, people thought of the God particle and the first push, believing that the ultimate source of all matter and energy is from God.

The first thing in this book that surprises the world is the cosmic force theory, CFT. Everything has no other source of mass and energy but comes entirely from the particles of which it is composed. Free particles are photons that travel at the speed of light and are very powerful — it goes without saying that the thing that flies at the highest speed is the most powerful thing in the universe. When a free particle encounters an object, part of its energy will be absorbed by the object, add its mass; the other part will be bounced off, add energy to accelerate the motion of the object — that's it.

The key to understanding this point is not to regard particles and energy as two things with different properties. They are both energy and the only difference lies in the speed of motion. The speed of light is a dividing line, which is also the dividing line between free particles & composite particles, energy & matter.

In addition to particles, atoms, and molecules, objects also have the tendency to aggregate. They are all based on particles, which are manifestations of the cosmic force. The most common and most powerful manifestation of the cosmic force is the spiral vortex motion. It forces everything to converge. They form the material force of the universe, and I include the cosmic spirit as part of the cosmic force.

The particles in MBT are powerless. Their motion depends on the external force with blurred faces and needs the guidance of more blurred intermediate particles.

This book highlights the characteristics of energy-mass continuum of particles, excluding all intermediaries and external forces. **Cosmic force is the solely source of all things,** and the whole process of the evolution of the universe is the process of particles, namely various varieties of photons, motion and their evolution. The creation, growth, movement and change of all things are simultaneously described as the movement and evolution of energy.

The Cosmic Spirit is an Absolute and Infinite Creativity

My religion consists of a humble admiration of the illimitable superior spirit who reveals himself in the slight details we are able to perceive with our frail and feeble mind.
Albert Einstein (Goodreads Quotes)

The material force of the universe is embodied in the substance of all things, but the cosmic spirit is embodied in the structure, relationships, and motions of all things. This makes the universe harmonious and perfect. **Man is the most exquisite and magnificent product of the cosmic spirit.** Only we can see the material force, master it, and use it to create numerous products that the universe has failed to create. On top of material products, there are more magical spiritual products, CFC is the latest one.

Cosmology is a new subject in science, but it has a philosophical origin. All kinds of sciences have their own philosophical origins. Plato's *Timaeus* and Aristotle's *On the Soul* and *On the Heavens* reflect the early cosmology, which can be called cosmology of the soul of the universe, or simply cosmology of the soul. In the sense of the term, soul is interlinked with mind and spirit. In Plato's and Aristotle's works, the soul is not only the source of the universe but also its driving force. Giordano Bruno, Copernicus, Johannes Kepler, and René Descartes are the inheritors and promoters of this theory. After reading Helen Dukas' book *Albert Einstein: The Human Side*, it's not hard to see that Einstein had a similar view.

Plato described the process of the cosmos as one in which Creator first put intelligent into the soul, and from the soul comes the physical universe, and the intelligent is given to everything in this way, and the world becomes orderly and produces all good things.

"Whereas He made the soul in origin and excellence prior to and older than the body, to be the ruler and mistress, of whom the body was to be the subject." (*Timaeus*, 35a)

"Now when the Creator had framed the soul according to his will, He formed within her the corporeal universe, and brought the two together, and united them centre to centre. The soul, interfused everywhere from the centre to the circumference of heaven, of which also she is the external envelopment, herself turning in herself, began a divine beginning of never ceasing and rational life enduring throughout all time. **The body of heaven is visible, but the soul is invisible, and partakes of reason and harmony, and being made by the best of intellectual and everlasting natures, is the best of things created.**" (*Timaeus*, 37e)

From the above two paragraphs, some people may regard *Timaeus* as the oldest Creationism treatise ever written. I'm not sure Plato believes in God, but Socrates undoubtedly believes in God. Supporters of intelligent design (ID) believe that there are aspects of the natural world that are so intricate and fit for purpose that they must come from the hand of an intelligent designer, and cannot be the work of 'The Blind Watchmaker' (The title of a book written by Richard Dawkins)

ID is a form of Creationism, which is generally considered a religion rather than a science; The origin of everything cannot be explained by Darwinian theory of evolution, as I have demonstrated in my book of *FROM FISHING APE TO GENIUS* (2013), and certainly not by the work of 'The Blind Watchmaker'. I ascribe it to the **Necessity of the cosmic force,** the cosmic spirit. As for Necessity, I think no one understood it better than G. W. F. Hegel.

"If all the conditions are at hand, the fact (event) must be actual; and the fact itself is one of the conditions: for being in the first place only inner, it is in fact itself only presupposed. Developed actuality, as the coincident alternation of inner and outer, the alternation of their opposite motions combined into a single motion, is Necessity...The theory however which regards the world as determined through necessity and the belief in a divine providence are by no means mutually excluding points of view. The intellectual principle underlying the idea of divine providence will hereafter be shown to be the notion... In the creed of the ancients, as we know, **necessity figured as Destiny...In the ancient creeds not only men, but even gods, were represented as subject to destiny,** a destiny which we must conceive as necessity not unveiled, and thus as something wholly impersonal, selfless, and blind. On the other hand, the Christian God is God not known merely but also self- knowing; he is a personality not merely figured in our minds, but rather absolutely actual." (*Shorter Logic* §147)

The common God of Socrates and Plato is personified, and necessity is the absolute and eternal that they repeatedly emphasize, that is, conditions exist and will happen, like divine will. The philosopher transcends the scientist in that he is never satisfied with facts and phenomena, but pursues essence and eternity hidden in facts and phenomena, the Ideas. This is the most basic requirement for cosmological research. To take the most obvious view, the space is full of stars, which constantly go around each other in complex orbits. In our common sense, it is a very fragile and volatile system. What principles make it stable and harmonious? The sole answer is necessity. "That everything that happens, happens of necessity." (Democritus) The origin, structure, character of the stars, and the cosmic background field, determine their fate as they are, and cannot be otherwise. **Everything, motions, changes, all of which arise and exist**

with their own necessity. Exploring and determining this necessity is the ultimate end of philosophy and science, and also the end of this book. The idea of 'identification of reason with nature', proposed by Europeans in the 18th century (Kline, 1953, p.270), is one of them.

Note here, necessity includes not only order & harmony, but also disorder & chaos. Only the latter will disappear in the process of evolution, making order & harmony in general. For example, there are many celestial bodies in the solar system, mainly comets and asteroids, which have been making irregular movements. They will eventually fall on the sun or other planets, and this disorderly movements will be eliminated.

Necessity is the supreme law of the movement of the universe, which is everywhere seen in nature. But nature can't speak, nor teach, and needs people to use wisdom to discover it. I call it the cosmic spirit, is to anthropomorphize nature.

We can firmly believe that the emergence of human beings is also due to necessity, not contingency. Necessity makes intelligent to nature, so that man comes into being and becomes the highest intelligent creature in the universe. We advocate freedom of worship, but we might as well worship nature as we worship God, and do two things well. One is to cherish nature and make it more and more colorful. The other is to improve ourselves, to possess virtues worthy of the title of intelligent beings. The higher requirement is to recognize and apply the cosmic spirit, and to devote ourselves to the creative activities of beauty and goodness.

The cosmic spirit is an absolute and infinite resource, nature's bounty. Needn't to spend money or ask for permission from any authority, anyone who searches it will find it and become creative and promising. In history, those great philosophers, theologians, thinkers, artists and scientists are all conscious or unconscious promoters of the cosmic spirit, our role models — **you don't just worship them, you can transcend them. This is the profound meaning of absolute and infinite.**

Darkened Light Theory (DLT)

Light is not seen as it is in the beam, but as it is thence reflected to our eyes; for vision can be no other wise produced than by rays falling upon the eyes.
Sir Isaac Newton

Mere light is mere darkness.
– G. W. F. Hegel: *Shorter Logic*, §36

"DARK LIGHT — an invisible form of radiation arising from the passage of ordinary light through apparently opaque metal plates, and capable of producing shadowgraphs like those of the X rays — have already been noticed in these columns. M. Le Bon's experiments have been described before the French Academy of Sciences, a sufficient proof that they have been performed in good faith; but his conclusions have been usually looked upon with suspicion, men of science having generally thought that he has been deceived by the filtration of light through chinks in his plate holder, or by some similar action." (https://borderlandsciences.org/project/aura/ article/SciAm_on_Le Bon _Dark_Light.html)

Light, no light. Light is dark, which is invisible to the human eye, and by that way light preserves its own energy, real dark energy. Everything we see is the interaction of light and matter, their vibrations and resonance. Visible light is the part of the spectrum that can see (and interact with) the surface of the object most easily, and the result of this interaction (in the form of reflected light) is the part that interacts most easily with the human eye. Visible light itself is also invisible. That is the only answer to the question of why the night sky is so dark.

Even if we look directly at sunlight, starlight and all

luminaries, we only feel light through our senses and interact with light, we do not *see* light. Because in any case, as long as there is light around us, we cannot rule out the interaction between light and things, especially with our retina and skin.

In scientific research, human vision is divided into physiological and psychological. In our daily life we often use them in combination, emphasizing psychological vision, that is not only seeing, but also perception. If we completely exclude psychological vision and simply talk about physiological vision, the eyes cannot see light and everything. The actual meaning of seeing is just to open our eyes and let the retina receive light and interact with it. The light that we have not yet received is not only invisible but also imperceptible. Watching means prolonging the interaction time between light and retina to produce stronger psychological visual response.

American lunar astronauts have inadvertently provided new and powerful evidence for this theory in recent years: there are no stars visible in the sky, on the moon there are none at all. The reason is clear: there is no air on the moon, so there is no interaction between air and starlight. The human eye cannot see the faint starlight that makes up the 3K background radiation.

I equate light with cosmic energy, or pure energy, which exists in the form of a field and propagates in the form of wave. That is the reason why people cannot see light. Cosmic energy is invisible and will not disappear, but it will interact with things to trigger energy-mass conversion and change shape, that's what we really see—sense.

Material energy, such as the energy hidden in coal and gasoline, is also invisible. What we can see is their material carriers. Like cosmic energy, they tend to accelerate the moving speed of objects, interact with things, give off light for us to see, and then their corresponding material carrier will be consumed. When they do not interact with things, without light, the material carrier is preserved. Starlight is the largest and most common material energy in the universe. Stars radiate energy all the time, and a small part of them interact with things that we see as starlight. Most of the energy that flows out has no interaction, is hidden in space, unseen to us, and becomes the main body of dark energy.

This discovery can be compared with Nicolaus Copernicus' findings. He revealed the myth of the day sky, and I have revealed the myth of the night sky.

ARGUMENT. why is the sky dark at night? This is a question that has been debated for more than 200 years. It's called the Obers' Paradox. In *A Brief History of Time*, Hawking described it this way.

"The difficulty is that in an infinite static universe nearly every line of sight would end on the surface of a star. Thus one would expect that the whole sky would be as bright as the sun, even at night. Olbers' counterargument was that the light from distant stars would be dimmed by absorption by intervening matter." (p.8)

He disagreed with this interpretation and offered his own, but controversial either.

In *An Outline of Force Cosmology*, I propose DLT as the best solution to the paradox. I regard DLT as an original theory, which was presented in the seminar of Beijing Relativity Theory Research Federation, on October 31, 2015, for the first time, under the title: LIGHT NO LIGHT — THREE WORDS TELL THE TRUTH OF THE NIGHT SKY AND THE UNIVERSE. Then it was published in the *Journal of Scientific Chinese*, on January, 2016.

When I read *The Mathematical Principles of Natural Philosophy* recently, I am surprised to find that Newton has this idea 300 years ago, and he is the pioneer of DLT, and I just developed and improved it. I believe that if Heinrich Olbers, the 18th century German astronomer and physician, read this passage, he might not put forward the dark night sky paradox.

The original text is relatively long, and Newton put forward it when he was arguing for the phenomenon of comet tail.

"The first is the opinion of such as are yet unacquainted with optics; for the beams of the sun are seen in a darkened room only in consequence of the light that is reflected from them by the little particles of dust and smoke which are always flying about in the air; and, for that reason, in air impregnated with thick smoke, those beams appear with great brightness, and move the sense vigorously; in a yet finer air they appear more faint, and are less easily discerned; **but in the heavens, where there is no matter to reflect the light they can never be seen at all. Light is not seen as it is in the beam, but as it is thence reflected to our eyes; for vision can be no other wise produced than by rays falling upon the eyes;** and, therefore, there must be some reflecting matter in those parts where the tails of the comets are seen: for otherwise, since all the celestial spaces are equally illuminated by the sun's light, no part of the heavens could appear with more splendor than another." (*The Principia*, 510)

NOTE. Although DLT follows Newton's view that light is invisible, **there is a fundamental difference between DLT and Newton's optics in terms of the nature of light.** Above all, Newton asserted that light was a kind of corpuscle, which belonged to matter, whose motion undoubtedly needed external force, and abide by Newton's three laws of motion. DLT asserts that light is the cosmic energy and cosmic background field, which is not only Self-motion, but also the total source of all energy & matter, I dub it the **Cosmic Force.** It obeys special laws of field motion, such as light velocity limit, energy - enhancing effect and light barrier effect.

Light Ball Theory (LBT)

Albert Einstein arrived at the conclusion that the maximum attainable speed in the universe is that of light. However, the **statement that claims stars, and all the luminous things that make light balls, keep expanding at twice the speed of light is also true.**

Take the Sun as an example. It sends light waves to all directions and makes light balls continuously. Its diameter is 2 million kilometres (600,000 km + 1,400,000 km of the Sun's diameter) in the first second. It will inflate to 37.4 million kilometres at 1 minute (600,000 km × 60 + 1,400,000 km) and 361.4 million kilometres at 10 minutes (600,000 km × 60 × 10 + 1,400,000 km), which exceeds the diameter of the spherical space between the Sun and the Earth.

This constant expansion of the Sun has been going on for at least five billion years. Currently, the solar bubble is a giant light ball with a diameter of ten billion light-years, and it continues to expand into outer space at the speed of six hundred thousand kilometres per second.

Space is full of light balls, big and small. They intersect and overlap each other to form a unified cosmic light field. This is the true cosmic microwave background source, but this does not overturn Einstein's theory of the limit of light speed.

Considering the edge luminaries, the universe as a whole should be a giant sphere of light with a diameter of more than one hundred billion light-years and should continue to expand infinitely outward at twice the speed of light. But only the sphere expands. The cosmic entity does not expand, instead it's contracting or holding steady.

In November 2018, NASA announced that Voyager 2 has finally crossed the edge of our Solar System through a giant **wall of fire,** which is made up of material ejected by our Sun and reaches temperatures of nearly 50,000 °C (90,000 °F). That hydrogen wall is the outer boundary of our home system, the place where our Sun's bubble of solar wind ends and where a mass of interstellar matter too small to bust through that wind builds up, pressing inward. It is impossible to block all sunlight from spreading outward, and at least half or most of them are trapped there.

Corollary, the radius of this spherical **'fire wall'**, is the radius of the core part of the solar ball, which is about 17 light-hours. The leaked sunlight that continues to spread outward constitutes the dim outer layer of the solar bubble. The whole 'fire wall' is actually a big hollow ball, which could be considered more of a fire shell or hot shell, with a diameter of 240 AU that wraps around the Solar System.

NOTE. When I conceived LBT, I consulted Stephen Hawking's description in *A Brief History of Time*. The original text is as follows.

"As we have seen, Maxwell's equations predicted that the speed of light should be the same whatever **the speed of the source, and this has been confirmed by accurate measurements.**

It follows from this that if a pulse of light is emitted at a particular time at a particular point in space, then **as time goes on it will spread out as a sphere of light whose size and position are independent of the speed of the source.** After one millionth of a second the light will have spread out to form a sphere with a radius of 300 meters; after two millionths of a second, the radius will be 600 meters; and so on. It will be like the ripples that spread out on the surface of a pond when a stone is thrown in. The ripples spread out as a circle that gets bigger as time goes on." (p.28-29)

Light Field Theory (LFT) ➤ Light Ether

The space filled with light balls is the light field. The two cannot be separated, nor can they be mixed together. The light balls originate from luminous objects and are specific, individual, and huge in quantity. **The light field is composed of all the light balls.** The specificity and individuality of each light ball disappears in the light field and is unified. In short, there are many light balls and only one light field.

However, light radiates outward from the source at the speed of light at all times, and the energy distribution of the light field is uneven, with relatively dense and relatively sparse regions, which will be discussed separately as local fields in this book.

We obtain the properties of the light field from the characteristics of the light ball, as follows:

1. Its source is matter. Without luminous substance, there is no light balls and therefore no light field.
2. The light field overlaps with material space, and the matter radiation in space forms 3K background radiation.
3. The nature of the field is energy, not matter, not made of corpuscles or chunks.
4. The light field is continuous rather than intermittent, that makes it a whole. Therefore, we cannot identify the speed of light as the velocity of a photon, but as the velocity of the light field.
5. The field of light diverges at the speed of light rather than converging, so it cannot be preserved. Note: if it could preserve all energy, it could also accumulate energy, which would disrupt the order of the universe, causing it to oscillate violently and perish.

6. The light field exists in the widest space but it does not occupy space exclusively. Energy encounters things that must interact with them and become part of them.
7. The light field is uneven — it is strong near sources of light, and weak when it is far away from them.
8. Light, no light; the field itself is not visible. What we can see is light's interactions with matter.
9. The light field transmits energy in the form of waves and has the effect of increasing energy. Thus, we can see the images of distant celestial bodies more clearly.
10. Light always radiates outward from the source; this is the direction of motion of light and light field. The cosmic light field is spreading outwards at twice the speed of light all the time.

Isaac Newton created gravitation theory and absolute time-space theory, on which later scientists built the theory of the gravitational field; Michael Faraday and James Clerk Maxwell founded electromagnetic field theory, and contemporary scientists created quantum field theory (QFT). **The light field is something almost nobody has paid attention to, it thus became my original theory. In fact, the light field envelops the whole universe and keeps expanding, and is the real cosmic background field.** The gravitational field, the electromagnetic field, and the quantum field are all in it.

The light field fills all the space and can transmit energy in the form of waves. Is it the ether, the quintessence, that people have been looking for for more than 2000 years? George Gamow, Russian-born American nuclear physicist and cosmologist, gave a positive answer. In the book, *One, Two, Three… Infinity: Facts & Speculations of Science*, he wrote.

"Since, however, it seems rather illogical to speak about something vibrating when there is nothing to vibrate, **physicists had to introduce a new notion, 'light-carrying ether,'** in order to furnish a substantive subject for the verb 'to vibrate' when attempting to explain the propagation of light...If we say that light consists of waves traveling through the light ether, defining 'light ether' as that through which light waves are traveling, we are telling a gospel truth, but also recording a most trivial tautology. It is an entirely different problem to find out what this light ether is and what its physical properties are. Here no grammar (not even Greek!) can help us, and the answer must come from the science of physics.

As we shall see in the course of the following discussion, the greatest mistake of the physics of the nineteenth century consisted in the assumption that this light ether has properties very similar to those of ordinary physical substances familiar to us. One used to speak about the fluidity, rigidity, various elastic properties, and even the internal friction of light ether. Thus, for example, the fact that light ether behaves on the one hand as a vibrating solid when carrying light waves, but on the other hand shows a perfect fluidity and a complete absence of any resistance to the motion of celestial bodies, was interpreted by comparing it with such materials as sealing wax. Sealing wax, and other similar substances, are, in fact, known to be quite hard and brittle in respect to forces acting rapidly in a mechanical impact, but will flow like honey under the force of their own weight if left alone for a sufficiently long time. Following this analogy, the old physics assumed that light ether, filling all interstellar space, acted as a hard solid in respect to very rapid distortions connected with the propagation of light, but behaved as a good liquid when the planets and stars, moving many thousand times slower than light, were pushing their way through it.

Such an anthropomorphical point of view, so to speak, which tried to ascribe to a completely unknown thing, which so far had nothing but the name, the properties of ordinary material known to us, failed very badly from the very beginning. And, in spite of many attempts, no reasonable mechanical interpretation of the properties of the mysterious carrier of light waves was found possible.

In the light of our present knowledge we can easily see wherein all attempts of that kind erred. In fact we know that all mechanical properties of ordinary substances can be traced back to the interaction between the atoms from which they are built...Light ether is a substance of a peculiar type, which has no similarity to the familiar atomic-mosaic that we usually call matter. **We can call light ether a 'substance'** (if only because it serves as a grammatical subject for the verb 'to vibrate'), but we can also call it 'space,' keeping in mind that, as we have seen before and will see again, space may possess certain morphological or structural features that make it a much more complicated thing than it is in the conceptions of Euclidian geometry. In fact, in modern physics the expressions 'light ether' (divested of its alleged mechanical properties) and 'physical space' are considered synonymous." (Gamow,1947, p.90-91)

Morris Kline, Professor Emeritus of Mathematics Courant Institute of Mathematical Sciences New York University, told us a fact that before Gamow, "Many of Maxwell's successors... The necessity for thinking in terms of a medium that carries electromagnetic waves soon established to their satisfaction the 'reality and substantiality of the luminiferous ether.' These pictures, however, cannot be taken seriously, for they are inadequate and experimentally non-verifiable." (Kline, 1953, p.318-319)

From *The Principia*: Introduction To The American Edition, we know that **Newton has given a more definitive answer to Ether more than 300 years ago, and it has been applied to his liquid vortex model.**

"In the paper entitled An Hypothesis Explaining Properties of Light, December, 1675, our author first introduced his opinions respecting **Ether—opinions which he afterward abandoned and again permanently resumed—'A most subtle spirit which pervades' all bodies, and is expanded through all the heavens.** It is electric, and almost, if not quite immeasurably elastic and rare." (*The Principia*, p.15)

Since Aristotle, most philosophers and scientists have agreed with this creed: The vacuum is not empty, and what permeates the whole space is an extremely thin thing, that is, the ether. It's just that no one can say clearly what ether is. Now, I can safely declare that **the light field is the light ether.** In which light not only flies unimpeded, but also its energy can be enhanced during flight.

NOTE. **The critical mass density of light field or ether is very low,** which has no any effect on the motion of objects. Michelson & Morley imagines it as a cosmic wind like air, and Newton imagines it as a liquid like water, which will produce friction and resistance to celestial bodies. They all exaggerate the density by hundreds of millions of times.

Furthermore. There is a self-evident truth that in space, **matter impedes motion and energy accelerates it.** That's why the speed of light can keep constant. In other words, the light field (ether) has no any resistance to any motion, only accelerates it, proportional to the speed of the moving object, that never faster than the speed of light.

Two Scientific Research Roads: Holism vs Corpuscularianism

For the sciences as a whole are nothing other than human wisdom, which always remains one and the same, however different the subjects to which it is applied, it being no more altered by them that sunlight is by the variety of the things it shines on. Hence there is no need to impose any restrictions on our mental powers; for the knowledge of one truth does not, like skill in one art, hinder us from discovering another; on the contrary it helps us.

René Descartes

"So many people today — and even professional scientists — seem to me like someone who has seen thousands of trees but has never seen a forest. A knowledge of the historic and philosophical background gives that kind of independence from prejudices of his generation from which most scientists are suffering. This independence created by philosophical insight is — in my opinion — **the mark of distinction between a mere artisan or specialist and a real seeker after truth.**" Einstein to Thornton (December 7, 1944, EA 61-574)

Holism and corpuscularianism come from the different research methods of philosophy and science. Philosophy is a comprehensive subject. Philosophical research tends to be holistic. Science is branch knowledge. Scientific research tends to be corpuscular theory. Philosophy and science are both Greek inventions, and the two theories, holism and corpuscularianism, were developed in the time of the ancient Greeks.

Corpuscularianism is similar to Leucippus and Democritus' atomism, except that where atoms were supposed to be indivisible, corpuscles could, in principle, be divided. It supposed all things to be composed of corpuscles. The theory became important in the seventeenth century; amongst the leading corpuscularians were René Descartes, Pierre Gassendi, Isaac Newton, Robert Boyle, and John Locke. Corpuscularianism later became the basis of phlogiston theory, particle theory, gene theory, mechanics and reductionism.

Holism, represented by Plato's theory of Form and the theory of soul, has not been developed because it is difficult to understand and operate. In fact, it died long ago, so corpuscular theory has become the mainstream theory.

The innate defect of corpuscularianism, however, is its one-sidedness. With the deepening and development of scientific research, this defect appears more and more obvious and becomes an obstacle to scientific advances. Field, energy, universe, and so on need to be based on the holistic theory in order to be in-depth and effective. Vortex motion is the most common and most powerful motion in space. It is not isolated but interrelated, interacting, and restrictive. It needs the guidance of holism in particular.

Whoever ignores holism will look at but not pay attention to or fail to understand the spiral structure of the sky, which is the new challenge facing cosmology.

Although Platonic philosophy died young, one of its unrivalled advantages is that it looks upward while other philosophies and sciences looking downward. If anyone knows this and applies it to the study of science, especially astronomy and cosmology, then such a person has earned the crown. In *The Republic* (529b) is the following passage:

"(The highest) Knowledge only which is of being and of the unseen can make the soul look upwards, and whether a man gapes at the heavens or blinks on the ground, seeking to learn some particular of sense, I would deny that he can learn, for nothing of that sort is matter of science; his soul is looking downwards, not upwards, whether his way to knowledge is by water or by land, whether he floats, or only lies on his back."

In *Laws* (903c, d, e), Plato offered the goal of looking upward: the **Whole** — the cosmic soul and the common good. He pointed out that you are created for the sake of the Whole, and not the Whole for the sake of you, and everyone does all things for the sake of the Whole, directing his effort towards the common good. This goal is also appropriate for contemporary people, and it is also needed by scientific research.

In the history of civilization, philosophy appeared much earlier than science, but it seems that there has been endless arguing around several core issues without substantial results, which is far less pragmatic and fruitful than science. Therefore, many scientists don't trust philosophy, and even oppose philosophy interfering with science. The title of chapter seven of Steven Weinberg's *Dreams of a Final Theory: The Search for the Fundamental Laws of Nature* is 'Against Philosophy', Einstein and Hawking have also expressed similar views.

They seem to ignore philosophical differences. But, leaving aside the other differences for the moment, closely related to science is the holism versus corpuscularianism, or deduction versus induction, which is discussed here. I assume that **scientists do not oppose induction because they have been practicing it and developing it all the time, whether they are aware of it or not. What they really reject is holism.**

Holism may not be suitable for all disciplines, but it is indispensable for cosmology. "The calculation of astronomical distances cannot be carried out by applying a yardstick. Moreover, whereas experience confines us to tiny portions of time and space, deductive reasoning may range over countless universes and aeons." (Kline, 1953, p. 27)

Scientific research cannot lack philosophy. In fact, everyone has some kind of philosophy to guide his mind and behaviors, but most of them are not aware of it. In philosophy, metaphysics belongs to holism, which is generally opposed by scientists. However, even Newton, an active advocate of experimental philosophy, could not get rid of metaphysics completely.

"Newton gave physics an express warning to beware of metaphysics, it is true, but to his honour be it said, he did not by any means obey his own warning. **The only mere physicists are the animals: they alone do not think: while man is a thinking being and a born metaphysician.** The real question is not whether we shall apply metaphysics, but whether our metaphysics are of the right kind: in other words, whether we are not, instead of the concrete logical Idea, adopting one-sided forms of thought, rigidly fixed by understanding, and making these the basis of our theoretical as well as our practical work. It is on this ground that one objects to the Atomic philosophy." (*Shorter Logic*, §98)

In Arthur Conan Doyle's works, Scotland Yard officials always paid attention to every detail of the scene, and build a case on them. What made private detective Sherlock Holmes superior to them was that he not only paid attention to details, but also was able to construct a **complete chain of details.** In this way, he could infer the whole event from a single link. "From a drop of water, a logician could infer the possibility of an Atlantic or a Niagara without having seen or heard of one or the other, So all life is a great chain, the nature of which is known whenever we are shown a single link of it." (*A Study In Scarlet*) Therefore, Holmes can avoid being deceived by superficial facts, and could find the missing rings of the chain, solved cases successfully.

Furthermore. Holmes could tell Dr. Watson's elder brother untidy habits and sad life from an old watch Watson handed him; Inferred the appearance and character of the user from a dirty old hat picked up in the street, and so on. This ability enabled him to vividly reproduce the characters and behaviors at various crime scenes.

In my opinion, this is also the practice of particle physicists. They waited in front of accelerators and colliders day and night, and discovered a large number of new and strange particles, thus inferring the nature and motion of everything, dating back to the origin of the universe 15 billion years ago.

Plato's theory of ideas tells us that truth is absolute and unique. Once we have acquired the highest knowledge, the eyes of our souls open wide to see ideals, we shall have insight into all things heaven and earth. This is the deductive science.

Just as Holmes was able to uncover the truth of events from microscopic traces, physicists should be able to discover the truth of the universe from particles with much greater ease. Because they're studying Nature, and **Nature never lies,** they can make repeated observations with state-of-the-art equipment. Holmes' objects are people, and it is much more difficult to reach the truth through a multitude of phenomena and illusions. Their differences stem from different beliefs. Scientists believe in and practice by inductive science, while Holmes, deductive science.

"First, **they came to consider deduction as the only respectable means of attaining knowledge**...Seduced by the success of the axioms in developing a system of geometry, the Greeks came to think of the axioms as '**absolute truths'** and to suppose that other branches of knowledge could be developed from similar 'absolute truths'...Now this worship of deduction from self-evident axioms was bound to wind up at the edge of a precipice, with no place to go. After the Greeks had worked out all the implications of the axioms, further important discoveries in mathematics or astronomy seemed out of the question." (Asimov, 1993)

It's true, if we overemphasize deduction we will make errs. But if we ignore it, we will make more and bigger errs, and even move in the wrong direction.

I have no doubt that if Scotland Yard had genetic technology invented by induction, Gregson or Lestrade, the two famous official detectives in *Holmes Adventures*, would have caught Jefferson Hope, the 'hero' of the tragic 'A Study In Scarlet', soon; but to convict professor Moriarty, they would have had to turn to Holmes. Because 'the Napoleon of crime' wouldn't leave his DNA or any other incriminating evidence at any crime scene. "He is the organizer of half that is evil and of nearly all that is undetected in this great city. He is a genius, a philosopher, an abstract thinker. He has a brain of the first order. He sits motionless, like a spider in the centre of its web, but that web has a thousand radiations, and he knows well every quiver of each of them. He does little himself. He only plans. But his agents are numerous and splendidly organized." (Arthur Conan Doyle. *The Final Problem*)

The classification of particles is also carried out independently by two different physicists. All particles are divided into two kinds by the name of their founder: bosons and fermions, but their connection is also suspicious. The most prominent is the electron, which is essentially an energy particle, but is classified as a fermion, a matter particle. This is the result of lack of deduction, or despise dialectics as F. Engels put it:

"Indeed, dialectics cannot be despised with impunity. However great one's contempt for all theoretical thought, nevertheless **one cannot bring two natural facts into relation with one another, or understand the connection existing between them,** without theoretical thought." (*Dialectics of Nature*, 1883)

Kline uses another words to express the same point, "Certainly one lesson to draw from the history of the Romans is that people who scorn the highly theoretical work of mathematicians and scientists and decry its uselessness are ignorant of the manner in which practical and important developments have arisen." (Kline, 1953, p87)

Dialectics, which I apply everywhere in this book, was what

Holmes called the science of deduction. a ready tool commonly applying in holism. This book may be regard as a unique deductive cosmology.

Cf. F.Engels: The Nature of Dialectics

"**The general nature of dialectics to be developed as the science of interconnections, in contrast to metaphysics.** Dialectics has proved from the results of our experience of nature so far that all polar opposites in general are determined by the mutual action of the two opposite poles on one another, that the separation and opposition of these poles exists only within their unity and inter-connection, and, conversely, that their inter-connection exists only in their separation and their unity only in their opposition." (*1883-Dialectics of Nature*)

"In the present work dialectics is conceived as the science of the most general laws of all motion. Therein is included that **their laws must be equally valid for motion in nature and human history and for the motion of thought.**" (Ibid)

Three Types of Corpuscular Theory

The holistic theory and the corpuscular theory are two ways of thinking. Though different in nature, they are not mutually exclusive. On the contrary, the two are interconnected and affect each other, and as methods they are also widely used. In contrast, the scope of application of corpuscular theory is much wider than that of holism, and holism appeals to corpuscular theory far more than the corpuscular theory appeals to holism. In this context, corpuscular theory has developed into three major categories.

The first, absolute corpuscular theory. This theory holds that **corpuscles have nothing, like geometrical dots. Everything about corpuscle is external, including mass, energy, motion, and mutual relations, all provided by messenger particles.** Even after that, if three quarks want to form protons, for protons and neutrons to then form a nucleon, they need messenger particles such as gluons or hadrons to paste or bind them together; like objects need gravitons to connect with each other, and so on. Even the repulsion of electrons requires the force of virtual photons. (Hawking, 2016, p.78) Previously, their mass and energy were provided by Higgs particles for the Big Bang. I call it mathematical corpuscular theory.

The second category is basically the same as the first, with one difference: messenger particles are produced by mass particles, they are not foreign, but they are members of the particle family. Their role as messengers may be full time or part time. Gluons, hadrons, leptons, π, W particles, Z particles, etc., are full time, while electrons, photons, virtual photon etc., are part time. Others believe that all particles are independent and can act as messenger particles at the same time. I call this physics corpuscular theory.

Hawking's description below expresses this idea very well.

"In quantum mechanics, the forces or interactions between mass particles are all supposed to be carried by particles of integer spin -0, 1, or 2. What happens is that a mass particle, such as an electron or a quark, emits a force- carrying particle. The recoil from this emission changes the velocity of the mass particle. The force-carrying particle then collides with another mass particle and is absorbed. This collision changes the velocity of the second particle, just as if there had been a force between the two mass particles." (2016, p.78)

The third category is something I came up with, called **the force corpuscular theory.** These kind of corpuscles are the energy-mass continuum. With reference to Indian physicist Satyendra Nath Bose and Italian physicist Enrico Fermi, **I classify all corpuscles into energy particles, mass particles & semi-mass-semi-energy particles.** Energy particles, Stephen Hawking called them 'force-carrying particles', are unconstrained free particles, flying at the speed of light, themselves are eternal energy, the motion of the power provided by themselves. **They are made up of photons and other particles traveling at the speed of light, including free electrons and neutrinos,** which are variants of photons. Mass particles refer to all composite particles, the most common being atoms. They make up molecules and objects, and they have energy in them, but most of their energy is trapped and cannot be used, only a small amount of spare energy can be used, so their velocities are relatively slow. That is to say, **both types of particles are composed of energy (photons), and both exist as energy fields, vibrating & transmitting as waves;** The only difference between them is that the former is a free state of energy with obvious characteristics of field and wave. The latter is a closed state of energy, its characteristics of the field and wave are not so obvious as the former, but the two are interrelated and mutual transformation.

In between are **the semi-mass-semi-energy ions, the most common and abundant particles that make up the star plasma, I call it the imprisoned bare proton**. It also moves at a speed in between, with the energy to vibrate, to move, and to glow provided by itself.

The main objects of physics are visible matter, developed from our daily life observations and experiences. Most objects that are observed, and the particles of which they are made, are individual and discrete, so the concept of corpuscles was popular in the ancient world, exemplified by Democritus' atomism. The phlogiston theory and the caloric theory, popular in Europe in the 17th century, were variations of this theory, on which Newton developed his corpuscular model of light. Although his theory was found to be flawed, it is still widely regarded and forms the basis of modern particle physics. Just as his law of gravity, which began with the problem of acting at a distance, has become the basis of modern field theory. Now when we're discussing about corpuscularianism, we can't ignore Newton's viewpoint. On the internet, I think the following two articles can explain the issue very well.

Just 100 years ago, A German-American astronomer John Martin Schaeberle wrote in one of his papers, 'In Newton's corpuscular theory of light' (https://edurev.in/studytube/Corpuscular-Theory-Class-11-Pysics), After other scholars collate and supplement, the following consensus is formed.

"**Newton's corpuscular theory of light** states that:
- **Light is made up of tiny particles called 'corpuscles'**, having negligible mass.
- These particles (corpuscles) are perfectly elastic.
- The corpuscles are emitted from the luminous sources such as Sun, candle, electric lamp etc.
- The tiny particles (corpuscles) always travel in a

straight line in all directions.
- Each particles (corpuscle) carries kinetic energy with it while moving.
- The corpuscles travel at high velocity.
- The corpuscul (light) would travel faster in the denser medium than in rarer medium.
  But latter this is proved wrong. We know that light travels faster in the rarer medium than in denser medium.
- when the particles (corpuscles) fall on the retina of the eye, they produce an image of the object or sensation of vision.
- The corpuscles can be of different sizes. The different colors of light are due to the different sizes of the corpuscles."

In recent years, three scientists from National Universityof Science and Technology of Zimbabwe — Golden Gadzirayi Nyambuya, Adrian Dube and Godwin Musosi – jointly published a paper on the internet, entitled 'Salvaging Newton's 313-Year-Old Corpuscular Theory of Light'. (https://vixra.org/abs/1710.0158) This is the abstract:

"As is well known — Newton's corpuscular model of light can explain the Law of Reflection and Snell's Law of Refraction. Sadly and regrettably — **its predictions about the speed of light in different mediums runs contrary to experience.** Because of this, Newton's theory of light was abandoned in favour of the wave theory. It [Newton's corpuscular model of light] predicts that the speed of light is larger in higher density mediums. This prediction was shown to be wrong by Foucault's 1850 landmarking experiment that brought down Newton's theory. The major assumption of Newton's corpuscular model of light is that the corpuscles of light have an attraction with the particle of the medium. When the converse is assumed, i.e., **the corpuscles of light are assumed to not have an**

attraction- effect, but a repulsion-effect with the particles of the medium, one obtains the correct predictions of the speed of light in denser mediums. This assumption of Newton's corpuscles repelling with the particles of the medium — this, might explain why light has the maximum speed in any given medium."

CF. Descartes' optical theory of light

"The principal properties of light are:
1) that it extends circularly in all directions around those bodies one calls luminous;
2) to any distance whatever;
3) instantaneously;
4) and ordinarily in straight lines, which should be taken as rays of light;
5) and that several of these rays coming from different points can collect together at the same point;
6) or, coming from the same point, can go out toward different points;
7) or, coming from different points and going to different points, can pass through the same point without impeding one another;
8) and that they can sometimes impede one another, namely when they are of very unequal force, that of some rays being far greater than that of others;
9) and, finally, that they can be diverted by reflection;
10) or by refraction;
11) and that their force can be increased;
12) or diminished by the different dispositions or qualities of the matter that receives them. Here are the principal qualities observed in light, and all of them are in accord with this action, as you shall see. " (Descartes, 2004, p.62-63)

The Quintessence of Holism: Light Field Theory & Vortex Theory

Scientists generally ignore holism, but they often encounter wholeness. The field subject is a typical one. Rodney A. Brooks, an Australian computer scientist, outlined field theories in his best-selling book *Fields of Color: The Theory That Escaped Einstein*, outlined three pillars of field theory.

"The field principle
The first pillar is the assumption that nature is made of fields. A field is a set of physical properties that exist at every point of space.

However, the concept of a field as a property of space does not come easily. It eluded the great Newton, even though he couldn't accept action-at-a-distance. It wasn't until 1845 that Faraday, inspired by patterns of iron filings, conceived the idea of fields, and it took another 50 years before the concept was accepted without invoking an imaginary ether.

QFT (Quantum Field Theory) comprises seven fields — five force fields and two matter fields. The force fields include gravity, electromagnetic forces, strong and weak nuclear forces, and the recently-discovered Higgs field. The matter fields include lepton and baryon fields. Two of these fields — strong and baryon — are effective fields that are made of more basic but 'invisible' fields called quarks and gluons. The use of colors is my attempt to make the field picture more palatable.

The relativity principle. The second pillar is the assumption that the field equations are the same for all uniformly-moving observers. This is Einstein's Principle of Relativity, famously enunciated in 1905. QFT is the only theory that successfully combines the relativity and quantum principles.

The quantum principle (discretization). The quantum principle was introduced in 1900 by Max Planck, who showed that EM radiation emitted by hot objects consists of discrete chunks that he called quanta. Discretization was demonstrated experimentally in 1922 by Otto Stern and Walther Gerlach." (2016, p.174-175)

Of the three field theories, the second is ambiguous. Brian Greene, a professor of physics and mathematics at Columbia University, wrote in his article, 'Special Relativity in a Nutshell', **"Einstein taught us that we must consider not only motion through space but motion through time. The two are inextricably intertwined."**
(https://www.pbs.org/wgbh/nova/article/)

But Einstein didn't tell us whether this interwoven space-time is a field. In an address entitled 'Ether and the Theory of Relativity' in 1920, He said clearly, "There can be no space nor any part of space without gravitational potentials; for these confer upon space its metrical qualities, without which it cannot be imagined at all. The existence of the gravitational field is inseparably bound up with the existence of space."

From this statement we know that **his field is an extension of Newton's gravitational field,** the only difference is in the background. Newton's gravitational field is in absolute space and time, they are not connected with each other; Einstein's gravitational field is in an interwoven space-time or curved space, which is interrelated and influences each other. In addition, Newton's equation of gravity has only one parameter, that has been confirmed by scientific experiments and has become the gravitational constant G. While Einstein's equation, according to Professor Timothy Ferris of the University of California, **requires as many as 17 input parameters, which were determined experimentally.** I quite agree with Einstein, "Any intelligent fool can make things bigger and more complex. It takes a touch of genius—and a lot of courage—to move in the opposite direction." (Goodreads-Quotes) With this belief, I don't trust formulas with more than three parameters, even though I have only middle-school math background.

The following passage in Brooks' book further confirms the internal connection between their theories.

"Where's the field
Astute readers will have noted that, for all of Newton's success, there was one thing missing. I'm referring to the subject of this chapter, the gravitational field. At the time, the discovery of the law of gravity was enough. It explained virtually all that was known in a way that everyone could understand, even if they couldn't do the math. That was enough to satisfy the most curious intellect and no one thought to ask how the force was transmitted. No one, that is, except Newton. This remarkable man, who had uncovered the secret of gravity, could not believe in action at a distance (see quote at the head of the chapter). He clearly wanted there to be a gravitational field, but finding no evidence he avoided speculation. **And so, scientists continued to believe that one body acts upon another at a distance for another 200 years, until Faraday and Maxwell developed the concept of the electromagnetic field. And it would be 50 more years before Einstein developed the field equations that gave gravity its 'full citizenship' in the family of fields."** (2016, p.37-38)

Paul Sutter, an astrophysicist at The Ohio State University and the chief scientist at COSI science center, wrote in his article, 'Einstein's Genius: Describing the Geometry of Space-Time': "The way that general relativity models gravity is through the dynamic machinations of space-time itself. According to the theory, the presence of matter and energy alters the fundamental space-time geometry surrounding those substances, and that altered geometry influences motion." (https://www.space.com/41416-einstein-relativity-geometry-space-time.htmlh)

From this, we know that Einstein's field theory is his space-time reference frame, which produces gravity and whose motion is influenced by matter and energy.

Since its inception, the theory of the light field has rightly become the representative work of holism. Because the boundary of the light field from the beginning far exceeds the boundary of the physical universe, and at twice the speed of light it expands forever. The electromagnetic field, the gravitational field, the quantum field, and so on, are all included in the light field, but they can only be local fields. The only way light can move is by radiation, which is an antigravitational force, and **the light field is an antigravitational field. Vortex motion is characterized by centripetal contraction, resulting in a smaller but more powerful vortex gravitational field. Together with the light field, they eliminates the effect of the distance, unites the universe, establishes the relationship between celestial bodies, and stipulates the order of the universe as a whole.**

All Fields Transmit Energy in Waves

As the small pebble stirs the peaceful lake,
The centre mov'd, a circle straight succeeds,
Another still, and still another spreads.
                                        Alexander Pope

Aggregation of matter also forms fields, such as the atmosphere, oceans, and spiral galaxies, called material fields. The difference between material fields & the light field is that main body of the former is matter —a large number of continuous matter — which is visible at a low speed. However, the main body of the light field is the light radiated by matter, which is high-speed and invisible. In comparison, the material field is limited and local, while the infinite light field is a cosmic field, but the principles of both are quite similar.

All moving bodies carry energy, mainly vibration, and this energy does not disappear from the field, even if the body is no longer in motion. The field then acts as a medium, transmitting and converting this energy. When an object enters the field, it produces an impact force that causes the field to react, blocking the transfer of energy and producing the first crest. The crest is the result of energy accumulation, and with the help of the field boosting its function it overcomes the reaction force and continues to push into the wider outer space, producing a second crest, which repeats the process and creates a third crest, which creates a fourth crest and so on, — this is the way the field transmits energy. Due to the great difference in power & velocity, the wave generated by the material field, including electromagnetic fields, has a low frequency and a long wavelength. The light field produces waves of high frequency and short wavelength.

We are now focusing on the energy field and mainly the light field. The characteristics of the light field are suitable for all the energy fields, and some are suitable for the material fields. They all transmit energy in wave form, which is common to all fields. Although each wave is independent & complete (Planck's 'chunks', 'quanta'?), and repeated successively, it is, on the whole, a field with waves that are continuous with each other. Waves only exist in fields, they cannot leave fields, always end-to-end. Intermittent signals are not called waves, but pulses.

Take throwing a pebble into water as an example. The pebble immediately sinks to the bottom of the pool and no longer moves, but the energy it carried diffuses across the water's surface in the form of waves, which is the way the energy field transfers energy. The astronomical phenomena we see is not physical — that is, photons emitted by the celestial body itself — but is energy waves transmitted by the light field. Because of the distance, most of them are very weak, which does no harm to organisms.

However, radiation is a common phenomenon. Objects that are close to us and radiate strongly will not only transmit energy waves to us through the field but will also directly transmit a material flow through space. The Sun is a celestial body, and the solar wind is a powerful ion flow; that's why sunlight does us harm and starlight doesn't. Water waves travel across the surface of the water. The image of a water wave is clearly visible, but we cannot trace the water wave back to the source object's image. What makes light waves special is that they are invisible, but we can trace them back to the image of the source object. I interpret this as the holographic nature of light waves, that is, each wave of light contains the whole (the full spectrum), and the whole can be seen through a pinhole, which is what the physical waves and other energy waves (electromagnetic waves) lack. The light field is composed of light waves, and the light field is characterized by continuity and integrity, so the light wave is holographic, and its efficiency is proportional to the amount of light. That's why, to get a clear picture of an object, astronomers need a large-aperture camera that's turned on for a long time.

"wave, propagation of disturbances from place to place in a regular and organized way. Most familiar are surface waves that travel on water, but sound, light, and the motion of subatomic particles all exhibit wavelike properties. In the simplest waves, the disturbance oscillates periodically (see periodic motion) with a fixed frequency and wavelength. Mechanical waves, such as sound, require a medium through which to travel, while electromagnetic waves (see electromagnetic radiation) do not require a medium and can be propagated through a vacuum. Propagation of a wave through a medium depends on the medium's properties.

Waves come in two kinds, longitudinal and transverse. Transverse waves are like those on water, with the surface going up and down, and longitudinal waves are like of those of sound, consisting of alternating compressions and rarefactions in a medium. The high point of a transverse wave is a called the crest, and the low point is called the trough. For longitudinal waves, the compressions and rarefactions are analogous to the crests and troughs of transverse waves. The distance between successive crests or troughs is called the wavelength. The height of a wave is the amplitude. How many crests or troughs pass a specific point during a unit of time is called the frequency. The velocity of a wave can be expressed as the wavelength multiplied by the frequency." (Encyclopedia Britannica)

The above description shows that the geometric characteristics of wave and vibration are basically the same, but their properties are different. **Waves are phenomena & matter; vibrations are forces & energy, and all waves originate from vibrations.** From this point of view, I paraphrase the title of this topic as:

All energy travels through the field (medium) in the form of vibrations, portraying as waves.

Enhanced Energy Effect of Field (EEEF)

The light field and all fields can not only transfer energy, but it can also enhance energy in the process of transmission. Its cause approximates the Fresnel drag effect, although they are not explained in exactly the same way. The light field is an energy ocean without boundaries. Only because reductionism is overwhelming, it is ignored. It should be said, if it was not through the light field and Earth's atmosphere, but through a vacuum, that the propagation of light depended only on radiation, the vast majority of stars would not be seen, and only the Sun and the Moon would be visible in the sky.

Cosmic rays are the most powerful proton radiation observed by human beings, though invisible. Isaac Asimov guessed that its source is "The most energetic one we can conceive of would be the mutual annihilation of heavy nuclei of matter and antimatter, and this would liberate at most 250 Bev." (Asimov, 1960, p.313) Well, I don't agree with the idea of antimatter. **I think most cosmic rays are the solar wind of the Milky Way enhanced by light field.**

Surface water forms a liquid material field, and the water field has the effect of increasing energy, especially in the initial stage, when the maximum peak of the water wave appears. We only need to consider that the outer ring of the water wave is larger than the inner ring, and the total energy is correspondingly larger, which is the manifestation of the energy-increasing effect. If you add up all the energy in the hydrosphere, you will find that it is many times higher than the energy carried by the pebbles, which is also evidence. But it has an obvious resistance, which is the air pressure & gravity of the earth, and it continues to act, causing the waves to eventually disappear into the water.

The light field is invisible, so here I use a material field to describe it but be careful not to confuse their properties. The difference between the two fields is mainly in density and velocity. **The light field diverges outward at the speed of light, so the density is very low,** and the velocity reaches the limit. But it won't be exhausted, because the radiation of countless celestial bodies is constantly replenishing it with energy.

The material field includes electric field, magnetic field, gravitational field, atmosphere, ocean and celestial vortex, etc., **all of which are characterized by high density and low velocity**. These fields are composed of matter that is never separated, and they move at roughly the same speed as that of the matter that makes up the field. From our daily activities and experiences, we can find a law of EEEF, which I call **'synchronization law'**. For an object to gain energy in the field, its density and velocity need to converge with those of the field. In other words, it requires its motion to be synchronized with the waves of the field.

Taking the previous example of throwing a pebble into water, the pebble itself cannot gain energy because it is too dense and stops moving soon after it hits the water. But it pushes some of the water molecules around, and they get EEEF.

This law becomes an **'irrelevance theorem'** when applied to the relationship between different fields. Scientists have long proposed that many fields with different properties overlap in space but are irrelevant to each other, as a result of their differences in density and velocity. The most prominent are the electric field and magnetic field, which move in opposite directions but always overlap, so that scientists regard them as an integrated electromagnetic field, but the two are actually irrelevant.

According to CFC, the interior of an atom is also made up of overlapping fields, not the vast void of one or more electrons orbiting a single nucleon. Protons and neutrons are strong, dense fields that form the core of atoms. The electron is a thin weak field, wrapped in all strong fields, e.g. the core, like the atmosphere wraps around the earth—Note. Both of them are vortex motion. The two kinds of fields are irrelevant, thus, electrons can act as nucleon's shell.

The light field surrounds all material fields and overlaps with all fields, as it is their background field, but these fields are irrelevant for the same reason. Meteors can better explain the synchronization law and the irrelevance theorem.

Meteors are pieces of rock or metal that enter the atmosphere. They are so numerous that they are constantly roaming in the field of light. Their density is much higher than that of the light field, and their motion velocity is much slower than that of the light field. They are out of sync with the motion of the light field, and can not be energized in the movement, so they will not emit light and cannot be observed by us. Compared with the light field, the atmospheric field has high density and slow motion, which is close to the density and velocity of the meteor. After the meteor enters the atmosphere, it is easy to synchronize with the atmospheric fluctuations, obtain the EEEF, and expand into a big ball of fire. In the process, most of its mass is converted to energy and disappears into the field, which we usually interpret as a meteor rubbing against the air and burning up.

EEEF has the most obvious effect on electrons. As we know, electron was discovered by British physicist Joseph John Thomson, Hawking told us that Thomson had observed the EEEF of electrons in the electric field, but only made a different explanation.

"In Thomson's experiments with electrons, we saw that **he used an electric field to accelerate the electrons.** The energy that an electron gains from an electric field of one volt is what is known as an electron volt." (2016, p.74)

1⊕ + 1⊖ = 30,000

In the 1930s, American physicist Carl Anderson confirmed the existence of positrons in experiments. After that, negative protons and antineutrons were discovered, one after the other. People then began to believe British physicist Paul Dirac's prophecy: every basic particle has a corresponding antiparticle in nature.

It comes naturally to people that if there are antiparticles, there must be antimatter. There is a general consensus in the scientific community that the result of a collision of positive and negative substances is annihilation and energy generation. This is undoubtedly a phenomenon of mass-to-energy conversion. The experiment of positive and negative particle collisions has therefore been taken seriously and become a hot topic of scientific experiments in the twentieth century, producing many achievements.

However, Frank Wilczek, an American theoretical physicist and 2004 Nobel laureate in physics, introduced the experimental results found in the United States and Europe in his book *The Lightness of Being: Mass, Ether, and the Unification of Forces*. All of these results show that matter does not annihilate itself in collisions. On the contrary, a large number of new substances are produced, its ratio is 1: 30000.

"At the Large Electron-Positron Collider (LEP), which operated at the CERN laboratory near Geneva through the 1990s, electrons and positrons (antielectrons) were accelerated to velocities to about one part in a hundred billion (10^{-11}) of the speed of light. Speeding around in opposite directions, the particles smashed into each other, producing a lot of debris. Atypical collision might produce ten π mesons, a proton, and an antiproton. Now let's compare the total masses, before and after:

- electron + positron: 2×10^{-28} grams
- 10 pions + proton + antiproton: 6×10^{-24} grams.

What comes outweighs about thirty thousand times as much as what went in. Oops." (p.16)

"A typical outcome would be, you collide two protons at high energy, and out come three protons, an antineutron, and several π mesons. The total mass of the particles that come out is more than what went in... Things don't appear to get simpler. It's as if you smashed together two Granny Smith apples, and got three Granny Smiths, a Red Delicious, a cantaloupe, a dozen cherries, and a pair of zucchini!" (p.31)

There are energy (Pions) and matter (protons), which Wilczek regarded all as matter. From the point of view of energy-mass conversion, it is allowed. Again, we can view them all as energy. The corollary to this is that collisions between electrons, protons, and other particles produce more particles that are not necessarily positive and negative in nature. Nor can we forget the opposite: the law of conservation of baryons in particle physics, the law of conservation of charges and leptons, and the law of conservation of singular numbers in strong interactions. Oops.

In the appendix of his book, Wilczek explained this phenomenon. he wrote:

"In Chapter 2, we considered a dramatic violation of conservation of mass. An electron and a positron annihilate, and outcome a collection of particles whose total mass is 30000 times larger. Nevertheless, energy is conserved. The velocities of the initial electron and positron were very close to the speed of light. Therefore, according to the general mass–energy equation, their energy is very large—much larger than mc^2. The particles that emerge from the collision, although they are more massive, move a bit more slowly.

When you add up their energies, calculated using the mass–energy equation, the sum matches the total energy of the original electron and positron." (p.208)

I think this explanation is worth discussing as it goes against **the popular idea of annihilation: total extinction, nothing left.** Another popular idea is that, at the big bang, space was filled with positive–negative pairs, and if the result of their collision was not annihilation but more matter, the total amount of matter in the universe would be tens of thousands of times higher than the actual amount!

From the point of view of energy-mass reciprocity, the only possible answer is the EEEF. The law of coservation of energy is applicable to material field, but not to energy field, especially light field, where energy is infinite. We have reason to believe that Wilczek saw only a small part, namely, the part of energy that converted into matter, and most of the newly added energy has not been observed, pure energy is invisible.

Important Scientific Discoveries Results from Particle Colliders

Although I advocate holism, I don't neglect reductionism experiments. Based on The most reliable experimental results of particle collision, I made some discoveries from EEEF. The data I use here come from Stephen Hawking's & Frank Wilczek's works.

The hot topic of science in the 20th century was the particle, which was associated with the development of nuclear weapons. The upsurge of building cyclotrons and colliders and conducting collision experiments has been higher and higher, more and more intense, fruitful, attracted the attention of all scientists, also mine.

I first noticed Frank Wilczek's experimental results in his book, as mentioned in the previous subsection; then a photograph and its caption in Stephen Hawking's book: "FIGURE 5.2 A proton and an antiproton collide at high energy, producing a couple of almost free quarks." (p.84)

Their data comes from the most advanced and authoritative scientific research institutions, whose accuracy is beyond doubt. **Why does Wilczek see a significant increase in energy or matter (imprisoned energy) while Hawking see a decrease in energy?** That is, he should have seen six free quarks released from two protons, not just two free quarks as the photograph showed. In neither experiment did true annihilation appear. I don't think particle physicists or quantum physicists can answer this questions, for none of them has noticed these phenomena and declare his discovery yet.

For solving this problem, we need to discuss a concept of particle classification. Physicists have discovered hundreds of sorts of particles, and an Indian physicist Satyendra Nath Bose paid attention to the light kind of particles (leptons), called bosons. The consensus among physicists is that Bosons are fundamental subatomic particle that has integral spin. Examples of bosons include photon and gluons. Before Bose, An Italian physicist Enrico Fermi paid attention to heavy kind of particles (baryons), called fermions. Most contemporary physicists think any particle that has an odd half-integer (like 1/2, 3/2, and so forth) spin. Quark, electron, protons and neutrons, are regarded as fermions.

Contemporary scientists have combined their findings to suggest that nature consists of these two kinds of particles. Fermions are particles of matter that make up all physical objects, and bosons are force carriers that transfer energy. This is a kind of MBT. which means that mass and energy are carried by the two special particles respectively. fermions have no energy and cannot move. Their energy in motion are provided by bosons.

Seen from this perspective, these two collision outcomes are weird. Electrons and protons are both fermions, protons are 1836 times heavier than electrons. After they collide respectively, a pair of light fermions produce a bunch of fermions, including a pair of protons, as Wilczek saw; While a pair of heavy fermions produces only the same number of light fermions(quarks), as Hawking saw, and we can't help but sigh with Wilczek: Oops!

The method of grouping hundreds of particles into two broad groups belongs to the holistic method, a top-down approach, I applaud it. In this book, they are described as energy particles (free particles) & mass particles. Free particles travel at the speed of light, which is energy, equivalent to what physicists call bosons; The composite particle moving at low speed or at rest is matter, fermions. They are called mass-energy conversion and energy-mass conversion of particles in this book, which are happen frequently. It belongs to EBT, which means that the particle is an energy-mass continuum that has energy of its own.

There are many differences between MBT and EBT classifications, other differences aside, we will now focus on one difference: what is an electron. According to MBT, the atom is the smallest unit of matter, and it is made up of electrons, protons and neutrons, which are all fermions, and their spin characteristics confirm this.

EBT, however, suggests that **electrons trapped in atoms are indeed the building blocks of matter, can be called fermions; But when they are free from atoms and become free electrons, they will immediately regain the ability to fly at the speed of light and becomes bosons.** (NOTE. This is not the original meaning of boson, which is just a force carrier, has no velocity index.) If you evaluate the results of the two experiments under discussion in this light, it would make sense.

What Wilczek saw was a pair of free electron-positron collision, the scene of a energy-mass conversion. Enter synchronization law. **Free electrons are photons, which travel at the speed of light and must be synchronized with the light field. When they collide, a lot of extra substances can be generated by EEEF.** Asimov in his *Guide* conformed it in a different way, "The cyclotron would not work for electrons, because at the high velocities needed to make the electrons effective their increase in mass was too great."

And, what Hawking saw was a lower speed proton-antiproton collision, the scene of a mass-energy conversion. A proton is a composite particle, its energy is trapped, that can't be used, and it has to rely on the cyclotron to provide power for its flight, which limits its velocity and keeps it from ever reaching the speed of light. Accordingly, its motion cannot synchronize with the light field, creating no EEEF. Their collision merely releases two of their miniature energy fields. (The two light dots in the photograph provided by Hawking are two miniature energy fields instead of two quarks. A basic notion in this book: Quarks exist in a proton in the form of aggregated micro-field.)

NOTE. Wilczek provides not only the result of two electrons colliding to produce more electrons and other particles (p.16), but also the result of electrons-protons colliding (p.31), contrary to Hawking's result and inconsistent with another result in his book, "At SLAC people actually shot electrons at protons, and observed electrons emerging after they collided. The emerging electrons have less energy and momentum than when they started." (p.40) This is a collision between energy particles (electrons) & mass particles (protons). Logically, they should be combined into atoms, not solely weakened electrons. I call it **DEEF (decaying energy effect of field).**

Rodney A. Brooks also talked about this phenomenon when discussing Einstein's special theory of relativity.

"MASS INCREASES. The final paradox of relativity is the increase in mass due to motion. Mass increase has been observed experimentally in particle accelerators, with increases as great as 3000% for particles traveling at over 99.9% the speed of light. How can the mass of an object get bigger just because it is moving?" (2016, p.150)

He doesn't specify what kind of particles are used in the experiment. But from the above discussion, we can be sure that they are energy particles rather than mass particles, which can't be accelerated to the velocity of light, and will gain no EEEF. It also shows that **EEEF is produced not only by the collision of positive and negative particles, but simply by the acceleration of the energy particles.**

CONCLUSION

Particle physicists have been experimenting with particle collisions, and

First, it is two kinds of particles, energy particles, or free particles, and mass particles. **There are only three kinds of energy particles, namely free photons, free electrons and neutrinos, which are energy rather than particles.** They exist as fields, employ their eternal energy to travel at the speed of light. The rest particles are mass particles, also known as composite particles, depending on external energy, flying at a relatively low speed, or of objects that cannot move. The two kinds of particles interact and transform with each other.

Second, energy particles collision result in energy and matter proliferation. Simple collision of mass particles do not have this result, they may produce the same or little more amount of debris.

Third, the extra energy and matter produced by the energy particles collision come from EEEF, rather than from mass increases as Special Relativity claims.

Fourth, the particles involved in the collision only spin in different directions, and there is **no destructive antimatter** among them. Because in all the experiments, scientists have been able to observe mass and energy increase events, but never an annihilation event. In addition, some experiments have shown that **EEEF can be produced without any collisions, just by accelerating energy particles.**

Fifth, the main body of sunlight is energetic particles, which has relatively weak EEEF, it is the main source of 'extraterrestrial dust' falling on Earth; The bulk of the solar wind is mass particles (ions), and there is relatively strong EEEF.

NOTE. The mass particles arise EEEF in two cases.
1. Being accelerated to the speed of light by the light field. That's how the EEEF of cosmic rays come from.
2. Entering in the dense region of the light field. Large numbers of matter will generate out from light barrier effects. (Ref. Cosmic Field and Material Fields)

Antimatter without a Foothold

Stephen Hawking has warned us. "There could be whole antiworlds and antipeople made out of antiparticles. However, if you meet your antiself, don't shake hands! You would both vanish in a great flash of light." (2016, p.77)

I think this warning is unnecessary because even if the antiperson can overcome all the material barriers and appear in front of me from an antiworld made out of antimatter, as soon as he gets out of the capsule, even as soon as he opens the door of the capsule, he will disappear in an instant into the air, so how can I have a chance to shake hands with him? To say the least, his ship must also be antimatter. After touching the dust or starlight of our material cosmos, or the square ground, the capsule and its occupant would be annihilated.

This case can overthrow antimatter theory, starting with Carl Anderson's discovery.

"Anderson placed in the chamber a lead barrier about ¼ inch thick. He found that the cosmic radiation crossing the chamber after it came through the lead did make a curved track. But he also found something else. In their passage through the lead, the energetic cosmic rays knocked particles out of the lead atoms. **One of these particles made a track just like that of an electron. But it curved in the wrong direction! Same mass but opposite charge. There it was—Dirac's antielectron.** Anderson called his discovery the positron. It is an example of the secondary radiation produced by cosmic rays; but in 1963, it was found that positrons were included among the primary radiations as well." (Asimov, 1993)

This raises the question of why these positrons did not disappear into the material space during the long journey.

Even better, why didn't they annihilate in the fog of the cloud chamber but leave a long trail? According to the aforementioned phenomenon of positive and negative substance collision and accretion, we have reason to worry about the safety of the cloud chamber and the personnel in the laboratory.

Collisions occur frequently, even violently enough to ionize cloud chamber atomic matter, but annihilation never occurs, so we have to doubt the authenticity of antimatter. It seems that everything is an artificially directed drama that does not exist in nature. Just as a group of anti-men have

infiltrated our planet, they can live peacefully with human beings. Only when they are arranged on the stage to fight with human beings can their true faces be exposed. That is, the annihilation will occur.

Think again, when you and your antiself shake hands or fight each other on the stage, not only do you two not disappear, as observed by F. Wilczek, but there are 29,998 or 14,999 pairs of you and anti-your are created from your contact, so what's annihilation here?

Each particle collider clearly announces that there is no antimatter in the world at all, and the colliding particles inside the machine are not positive and negative pairs. Because to be truly positive and negative particles, they don't have to collide at high speed, any small contact — anyone gently shaking hands with one's antiself— will cause them to annihilate.

I really appreciate that someone called the collider 'a super toy in the field of high-energy physics'. In *The Force Cosmology* (2017, not successfully published), I cheered for the US Congress forced suspension of SSC (Superconducting Super Collider) in October 1993.

My book claims that matter is atoms. The atom that gained extra electrons becomes anion, and that lost some electrons becomes cation, and you could call cation matter and anion antimatter, thus supporting the antimatter view. But one obvious fact is that they don't annihilate when they meet. Instead, they quietly revert to normal atoms.

Ubiquitous Antiparticles

There is a saying that no two leaves in the world are the same. There is a generalization that there is no antimatter or antiobject in the universe. However, antiparticles are everywhere.

From Carl Anderson's discovery, we know that positrons and electrons have exactly the same mass and property, except that it bend in opposite directions in a magnetic field, that is, it carries the opposite charge. Other antiparticles such as antiprotons, antineutrons, and antiatoms are the same in nature as their bodies, except when it comes to the different directions of motion or spin. According to the spherical symmetry theory mentioned earlier, this difference is universal, with a large number of particle-antiparticle pairs at any given time made by the Sun. From this perspective, the pairs of positive and negative particles are moving in the same direction, but they are in opposite space. The previous discussion has made it clear that their collision will not result in annihilation but instead increase mass by dragging field energy. We can believe that not only the collision of particle-antiparticle pairs but also any collision of a particle pair will produce similar results.

That is to say, **all particles are antiparticles from the mirror image, so the so-called antiparticle scarcity theory has become a pseudo-proposition.** For example, why do we observe the photons of sunlight moving in the same direction but cannot see the photons moving in the opposite direction? The answer is very clear: the same number of photons moving in opposite directions are on the other side of the Sun, and we cannot observe both sides at the same time. We can replace the sun with a spherical photon (or proton) emitter and measure the radiation on both sides at the same time, and then we can measure the particles and antiparticles pairs easily. It's what quantum physicists call quantum entanglement.

This theory cannot be extended to the atomic and molecular levels because they are matter, while particles are a continuum of energy and mass. The properties and structures of these two things differ greatly and the laws applicable in the particle world have no universality in the material world. The extension of this theory to objects, such as an antihuman, an antiearth, and an antiuniverse, is purely a work of science fiction.

Conclusion: antiparticles are everywhere, and there is no antimatter. Note that the antiparticle here is not the kind predicted by Dirac. The antiparticle here is confined to the direction; Dirac imagines antiparticle that not only spin in the opposite direction, but annihilate when they meet.

Not only are antiparticles everywhere, but so are antiforces. **Every force causes an equal antiforce**, and they come in pairs, based on Newton's third law of motion. In this book, there are a lots of force-antiforce pairings: aggregation-dispersion forces, vortex-explosive forces, gravity-antigravity forces, contraction-expansion forces, centripetal-centrifugal forces, pressure-tension, etc. It should be noted that dispersion force, explosive force, expansion force, centrifugal force, tension and radiation, are all classified as antigravity forces. This is not innovative, but not many people explicitly advocate antigravity because it would shake up Newton's law of gravity and standard cosmology. In fact, antigravity is everywhere and powerful.

Optical Image

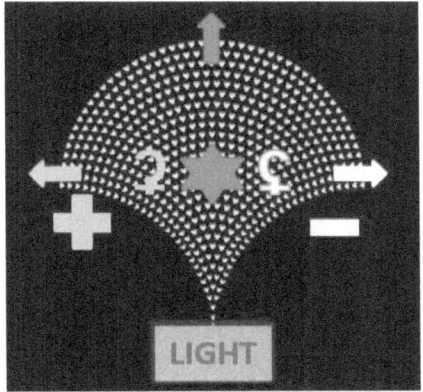

Obvious symmetry ∽ mysterious entanglement

New Interpretation of the Wave–Particle Duality of Light

"The concept of wave-particle duality was introduced by Einstein in the 1905 paper that earned him the Nobel Prize. He argued that, since EM radiation is emitted in discrete units by single atoms (as Planck had shown) and is absorbed in discrete units, then surely each unit must be localized in space—like a particle. How else could it be in a position to deposit all its energy into a single atom? On the other hand, the validity of Maxwell's equations for the EM field was indisputable and the wave behavior of EM radiation was well-documented. Thus was born the idea of wave-particle duality.

This concept was extended to matter in 1920 by Louis de Broglie, who showed that **the electron, long thought of as a particle, also exhibits wave characteristics** (see Chapter 6). If the photon's particle-like behavior could not be ignored, then the electron's wave-like behavior was even less ignorable. And so Schrödinger's famous equation was not interpreted as a field equation as Schrödinger would have liked, but as an equation that gives the probability of finding a particle at a particular location.

Resolution. The wave-particle duality paradox is resolved in a very simple way. In QFT there are no particles; there are only fields. **The explanation for the particle-like behavior is quantum collapse,** as described in Chapters 3 and 6. Each field quantum has its own identity and acts as a unit. If a quantum is absorbed by an atom, all its energy is deposited into that atom, no matter how spread out it may be." (Brooks, 2016, p. 166)

Is light a particle or a wave? This proposition comes from the opposition between Newton and Huygens. In the 20th century, Einstein's photoelectric experiment confirmed the particle behavior of photon and electron, while French physicist Louis Victor de Broglie and the German physicist Erwin Schrödinger continued the debate by proposing wave-behavior and formulas for electrons. The solution of quantum theory is to replace photons with quanta and eliminate particles with fields. In a word: there are only fields, no particles.

However, we can't deny particles. Our body and things we saw are all composed of atoms, molecules, ions and other particles. We can't deny waves either, because we live in light waves, sound waves, radio waves or water waves. It's indeed a paradox that photons cannot be both waves and particles at the same time.

The fact is that the photon is switching back and forth between the two states, like a man with invisibility, but at the speed of light, which is much faster than man can feel or imagine, thus creating the illusion that the photon is both a particle and a wave. The idea of eliminating the paradox is as follows.

Confirm that **the universe is made up of energy & matter. Energy is the source and matter is its product.**

The basic form of energy is the field, which is also the background of the universe. The field is unified, invisible, constantly expanding outward as waves at the speed of light, interacting with all matter.

Matter is made up of particles, most of which are visible and move slowly. The agglomeration states of matter form various material fields. Particles are imprisoned energy, miniature fields embedded in the field of matter, which in turn is embedded in the field of cosmic energy.

Energy produces matter through energy-mass conversion, and matter returns to energy field through mass- energy conversion. These two kinds of conversion movements occur at all times, that is, the wave form of photon and particle form are exchanged at all times.

Free photons (and free electrons) are energy, have no mass, travel as waves with a velocity of C. When they encounter matter, they immediately break off their flight and interact with matter, becoming part of matter, showing particle behavior, moving at a low speed or at rest.

Light itself does not emit light, only waves and fields, no any particles. When the light comes into contact with matter including the measuring instrument and our eyes, the interaction occurs, and the light is immediately converted into imprisoned photons, or particles, or quarks. **We can only see and observe photons and particles (no one has ever seen a quark yet), and never see the light and light field behind them.**

Photons, though very active, alternating between motion and rest, make up only a small part of the vast light field. Their behavior has no effect on the light field, but the light field's effect on them is constant and powerful, because photons cannot be separated from the light field; they are always in the field. This requires us to observe any motion of photons in relation to the light field. Just as we can't ignore water when studying the movement of fish. Ignoring the light field, it is impossible to understand the various properties and phenomena of light accurately and profoundly. The discussion of the subsequent double-slit experiment is an example.

All Particles Are Derived from Photons and Live Forever

That which always was, and is, and will be everliving fire, the same for all, the cosmos, made neither by god or man, replenishes in measure as it burns away.
Heraclitus, *Fragments*

The development process of everything and the universe is from simple to complex, from low to high. I put forward this proposition accordingly. **The original source of all particles is the photon, but the transformation forms of particles are various. Thus we return to Heraclitus, whose 'fire' is to be regarded as light; it does not need the first push.** Fire is his first principle; all things are exchanged for fire and fire for all things. Fire changes into various other elements in a cosmic cycle.

One of scientists' visual illusions is to regard the transformed forms of particles as new particles and strange particles. So far, scientists have discovered hundreds of particles, and most of them come from this regard.

Particles are the force of the universe. They can only transform, never die, and eternal vibration and resonance are characteristics common to all particles.

It is another misunderstanding when contemporary scientists conclude that the life span of particles is short. Atoms and the particles below them are invisible. What we can see is the interaction between particles or between particles & objects. That is just the state of particle transformation. This interval is very short. Scientists have mistaken it as the lifetime of particles. Among them, hundreds of resonant particles have the shortest lifetime.

The origin of photons is clear—they all come from luminous celestial bodies and burning objects—but the origin of quarks, electrons, mesons, neutrinos, etc., is not clear, which is also the basis for this proposition.

Light exists in the form of a field and is invisible. It is only seen and behaves as particles when it interacts with objects, which is also the case with particles. Without interaction, there would be no particles and no particle fields, the same is photons & light field.

Two forms of light-object interaction are the most telling.

One is the spectrum produced when light passes through a prism, the other is the rainbow produced when light passes through a drop of water. In these instances, the properties of the photon are the same, the properties of the prism molecules are the same, and the properties of the water molecules are the same, so that we will not mistake the different light (presented as colors) generated by a triangular prism and a water drop as photons with different properties.

With this paradigm, we won't hold quarks, electronics, neutrinos, mesons, hadrons, leptons, and so on, as species with different sources and different properties, but as members of a large family of photons, which show various differences only because of their interaction with things or because of different environments. Nor do we think of them as antiparticles as they differ in forms, direction of spinning, positive and negative charges. Just like people are divided into men and women, among these two gendered groups there are left-handed people and drunks, but no one becomes antihuman for these differences — they shake hands and hug each other every day, but never worry about mutual annihilation.

We can believe that **light is holographic and that the light field is everywhere,** and no matter where it is, a light spot as big as the tip of a needle will appear as a complete spectral band through the prism — this is the product of the light field. If this light spot is irradiated on the human body, what appears to be a point of light is actually a complete series of lights interacting with our bodies. It has the longest red wave, which is intercepted on our body surface and interacts with epidermal molecules to make us feel slightly warm and form leptons. Violet light goes deep under the skin, destroying atoms in our bodies or combining with particles to form baryons. In this way, light is equivalently converted into matter, which is mostly damage to normal people. The shortest ultraviolet light will pass through our bodies, and even the Earth, from the gap between molecules and atoms, and is considered to be the most abundant neutrino in the universe. Now, if we expand this light spot into the Sun, all kinds of particles are produced from this kind of interaction, as there are countless suns in space and when we intercept their lights with instruments, we can get an average value, which is background radiation.

What a photon looks like?

A photon is the smallest unit that scientists and their counter can determine when light field interacts with matter, including the counter. **It is actually a light wave, which won't really break away from the light field under any circumstances.** We usually say that light has seven colors, which are photons with different wavelengths and frequencies, but with **constant velocity, that is, $c = \lambda f$.**

Here is a question to be determined: Does each photon (wave) contain seven colors, or does each of the seven photons have unique color? I agree with the first statement. Because light is holographic, and colors are just light waves of different wavelengths and frequencies, far more than seven kinds, we can break it down into dozens or more, and each photon (a wave) actually contains the whole spectrum. An evidence: the spectrum is a whole, and scientists have not yet succeeded in separating or making any monochromatic light from it.

After Planck established quantum theory, Einstein put forward the concept of light quantum. Photon and quantum are regarded as particles with the same properties. On this basis, Planck established Planck constant, which is a unit of energy (quanta) pertinent to microscopic systems for which quantum mechanics applies. It helps us understand photons. It's just, we have to keep in mind that light is a field, inseparable. Photons are just physical phenomena that we see when we measure light.

SUMMARY. Everything is different forms of light, and the universe is the arena of light evolution.

Cf. Nikola Tesla on Light, Electricity, and Others

Recently, I am surprised to find that my idea is not new a hundred years ago, when the Serbian-American physicist engineer and inventor **Nikola Tesla** whom, I call him modern Heraclitus, announced in an interview with journalist John Smith in 1899: **Everything is the light**. In that interview, he also comments on electricity, vacuum, ether, and relativity, all of which are surprisingly close to mine. "Part of this interview is dedicated to Tesla's critics on Einstein's theory of relativity that discards the ether is energy. I have proved in the new Theory of the Universal Law why Einstein's theory of relativity is entirely wrong and why there is no vacuum (void), and that everything is energy. Thus I confirm Tesla's ideas as expressed in this interview." (J. Smith)

The following is an excerpt from the original text of electrical-engineering-portal.com, September 12th, 2012.

"I know that gravity is prone to everything you need to fly and my intention is not to make flying devices (aircraft or missiles), but teach individual to regain consciousness on his own wings … Further; I am trying to awake the energy contained in the air. There are the main sources of energy. What is considered as empty space is just a manifestation of matter that is not awakened. **No empty space on this planet, nor in the Universe. In black holes,** what astronomers talk about, **are the most powerful sources of energy and life**."

"**First was energy, then matter.**

Exactly! What about the birth of the Universe? **Matter is created from the original and eternal energy that we know as Light.** It shone, and there have been appear star, the planets, man, and everything on the Earth and in the Universe. **Matter is an expression of infinite forms of Light, because energy is older than it.** There are four laws of Creation. The first is that the source of all the baffling, dark plot that the mind cannot conceive, or mathematics measure. In that plot fit the whole Universe. The second law is **spreading a darkness, which is the true nature of Light**, from the inexplicable and it's transformed into the Light. The third law is the necessity of the **Light to become a matter of Light.** The fourth law is: **no beginning and no end**; three previous laws always take place and the **Creation is eternal.**"

"**Remember, it is not curved space, but the human mind which cannot comprehend infinity and eternity!** If relativity has been clearly understood by its Creator, he would gain immortality, even yet physically, if he is pleased. **I am part of a light, and it is the music. The Light fills my six senses: I see it, hear, feel, smell, touch and think.** Thinking of it means my sixth sense...No, I have nothing against Mr. Einstein. He is a kind person and has done many good things, some of which will become part of the music. I will write to him and try to explain that **the ether exists, and that its particles are what keep the Universe in harmony, and the life in eternity.**"

"**Everything is Electricity. First was the light**, endless source from which points out material and distribute it in all forms that represent the Universe and the Earth with all its aspects of life. **Black is the true face of Light, only we do not see this.** It is remarkable grace to man and other creatures. One of its particles possesses light, thermal, nuclear, radiation, chemical, mechanical and an unidentified energy. **It has the power to run the Earth with its orbit. It is true Archimedean lever.**"

"Why would you need to understand, Mr. Smith? **Suffice it to believe it. Everything is light.** In one its ray is the fate of nations, each nation has its own ray in what great light source we see as the sun. And remember: no one who was there did not die. They transformed into the light, and as such exist still. **The secret lies in the fact that the light particles restore their original state.**"

"**I prefer to call it: return to a previous energy. Christ and several others knew the secret. I am searching how to preserve human energy. It is forms of Light, sometimes straight like heavenly light. I have not looked for it for my own sake, but for the good of all. I believe that my discoveries make people's lives easier and more bearable, and channel them to spirituality and morality.**"

STATEMENT. FIRST CAUSE: LIGHT

The phrase of "Everything is light" is actually "Light creates everything." The creator of all things is not God, nor man, but Light. It is put forward independently as an assertion, not as a hypothesis, by three thinkers of different ages— I am one of them.

"Viewing fire as the essential material uniting all things, Heraclitus wrote that the world order is an 'ever-living fire kindling in measures and being extinguished in measures.' He extended the manifestations of fire to include not only fuel, flame, and smoke but also the ether in the upper atmosphere." (Encyclopedia Britannica)

Accordingly, Heraclitus is the first advocate of 'Light Creator', and Tesla is the second one.
 I am the third advocate, the first expositor, and apply it to establish CFC, thus develop this assertion into the highest knowledge of 21st century.

The Speed of Light is the Boundary Between Matter and Energy

Einstein's theory and formula of mass-to-energy conversion are both generally accepted. In *Asimov's New Guide to Science*, Asimov used the formula E = mc² to show us the theory. He first determined the speed of light, c = 3 × 10^{10}cm/s, to the square of the speed of light, c² = 9 × 10^{20}cm/s, and converted it into the smallest unit of energy, an erg, to get the result that "the energy in one gram of mass is sufficient to keep a 1,000-watt electric-light bulb running for 2,850 years. Or, to put it another way, the complete conversion of a gram of mass into energy would yield as much as the burning of 2,000 tons of gasoline." (Asimov, 1960, p.367)

The necessary condition for a particle to convert to energy is a velocity C, which only energy particles can achieve. Mass particle, e.g., atoms, molecules and ordinary objects cannot produce this result, for they are composite objects, and their internal particles move in different directions; In addition, their own energy is trapped, so they have to rely on external energy for motion and flight. Anything that depends on external energy cannot move at C, only energetic particles, viz. free photons, free electrons, neutrinos, can, because they themselves are energy. The boundary between matter and energy lies at C. If particles include energy particles move slower than C, even if it is very close to C, it is matter, not energy. Once its speed of motion reaches C, it will be no longer be matter, but energy. In other words, as long as the particle has mass, it cannot be moving at C; only massless particles travel at C. The speed of light is the limit of speed, the symbol of cosmic energy, and the end of matter.

A good exponent of it is electron. A free electron has no mass, and when it is imprisoned in an atom, which is moving according to the Rutherford model of the atom, just a little below the speed of light. It transforms from pure energy into energy particle, **gaining a tiny mass, one 1,836 times the mass of a proton.**

This confirms Einstein's assertion that nothing can go faster than light. We may then query all phenomena that are claimed to be superluminal, except the inflating motion of light ball and of light field described in this book, which don't violate Einstein's assertion.

However, the particle does not disappear from the universe in this case. It enters the field and becomes a part of it. Particles incorporated into the field are no longer matter but energy, and the difference among them disappears. The relation of fields to particles is analogous to that of oceans to raindrops. Although raindrops (including hails and snowflakes) differ from each other in shape, mass and motions, they are more different from the ocean than they are from each other, and yet all of these differences disappear when they submerge into the ocean. We can also subdivide raindrops into molecules or atoms or particles, and the analogy is even more relevant.

There are three differences between particle physics (MBT) and EBT. First, mass-energy conversion. **EBT holds that the mass of a particle is its original amount at rest, decreasing as the speed of motion increases, which is converted into energy. At the speed of light its mass is zero, that has completely converted into energy, and assimilate into the light field.** Particle physics holds that the mass of a particle is zero at rest, and mass increases with the speed of motion, reaching an infinite mass at the speed of light; in the same, its length is contracting to zero.

Second, the object of M. As for M, **EBT determines M in the formula as energetic particles, which is not applicable to other things** including mass particles and trapped photons. In Einstein and Asimov's case, M refers to everything, including all particles, atoms, molecules, objects, and people. On this basis, Einstein and other scientists have constructed the scaling effect, time-space travel, twin paradox, wormhole theory, etc.

I hold a negative attitude toward these. Atoms, molecules, rulers, people, and other objects must be decomposed into undifferentiated energetic particles first to reach the speed of light, and the speed of a free photon is exactly C.

Third, the demarcation of matter. **EBT believes that the speed of light is the demarcation of matter. Once this speed is reached, matter disappears and is converted into energy, which is incorporated into the light field at once.**

MBT believes that matter is infinitely divisible, that is, there is no demarcation and matter is eternal. This view was established by Newton.

"and thence we conclude the least particles of all bodies to be also all extended, and hard and impenetrable, and moveable, and endowed with their proper vires inertia. And this is the foundation of all philosophy. Moreover, that the divided but contiguous particles of bodies may be separated from one another, is matter of observation; and, in the particles that remain undivided, our minds are able to distinguish yet lesser parts, as is mathematically demonstrated. But whether the parts so distinguished, and not yet divided, may, by the powers of Nature, be actually divided and separated from one an other, we cannot certainly determine. Yet, had we the proof of but one experiment that any undivided particle, in breaking a hard and solid body, suffered a division, we might by virtue of this rule conclude that **the undivided as well as the divided particles may be divided and actually separated to infinity.**" (*The Principia*, 389)

Mass-Energy Conversion: Changing Speed with Mass

The mass of a body is a measure of its energy content.
A. Einstein

Matter and energy can be equally converted to each other. This is the theory founded by Einstein. He also proposed a mathematical formula, $E = mc^2$. Energy is invisible, matter and motion are visible, and **we may as well equate energy with the speed of material motion, meaning energy becomes intuitive.** That's what Einstein's formula means: energy on the left-hand side of the equation, velocity on the right-hand side. The classic example is the scaling effect.

The effect is, in essence, that a ruler shrinks by 15% when its speed reaches 150,000 km per second (half the speed of light). Then, when reaching 260,000 km per second (7/8 the speed of light), it shrinks by 50%, and when reaching the speed of light (299,793 km/s), its length is zero, as all mass is converted into speed. Einstein came to the conclusion that in the theory of relativity, the meaning of velocity, C is the limit of velocity, and any real object can neither reach nor exceed it.

Here we need to consider the source of energy, that Einstein didn't provide the answer but Hawking did by increasing mass.

"For example, at 10 percent of the speed of light an object's mass is only 0.5 percent more than normal, while at 90 percent of the speed of light it would be more than twice its normal mass. **As an object approaches the speed of light, its mass rises ever more quickly, so it takes more and more energy to speed it up further.** It can in fact never reach the speed of light, because by then its mass would have become infinite, and by the equivalence of mass and energy, it would have taken an infinite amount of energy to get it there. For this reason, any normal object is forever confined by relativity to move at speeds slower than the speed of light. Only light, or other waves that have no intrinsic mass, can move at the speed of light." (2016, p.24)

This passage tells us that the high-speed motion of an object requires corresponding energy support and corresponding mass that provides the energy, but **it is not clear whether this mass is generated by the object in motion, or if it is external. In my opinion, it is impossible for an object to obtain mass when flying at high-speed and automatically convert that mass into energy, it can only come from its own mass- energy conversion to reach the speed of light.** Another consideration is also necessary; that is, that all particles originate from photons, and their inherent speed is the speed of light.

Any ruler is ultimately a combination of particles. Now we don't think of the ruler as a ruler, but as a combination of particles whose ability to move is restricted. The movement of the ruler is now converted into the movement of every particle. At this point, the particle motion becomes the anti-binding behaviour (remove shells) that reduces the mass of each particle and increases the energy of each particle. As a result, the movement speeds up and the overall volume expands rapidly. When reaching the speed of light, the binding force is completely eliminated. The movement of particles is now not subject to any restriction; their speed reaches the limit, and all mass particles turn into energy particles, all are free. We only catch sight of a flash and the ruler disappeared.

This description is logical, but it can't happen in the physical world. It is impossible for a ruler or any other ordinary object to shorten by a fraction in flight because the constraints on the motion of the particles of which it is made up are not removed. **The only way to produce a scaling effect is to break the ruler down into particles,** and the only way to do this is to increase the temperature. External energy sources have been excluded here, so warming means that particles are moving faster, **and their energy comes from unbinding itself.** They all go away. In other words, the density of the ruler drops to zero, but they don't disappear from the universe. They turn into invisible energy waves and enter the cosmic background field, with matter-burning and luminescence being the most common phenomena of mass-energy conversion. Again, it's hard to call it a scaling effect.

This discussion exposes a divergence in corpuscular theory. The popular view is that particles have no mass or energy, that they got their mass from the Higgs boson at the Big Bang, and that they combine and move by relying on four forces of nature. The four forces in turn come from massless sub-particles or bosons — such as gluons, muons, gravitons and most likely, the photons and virtual photons. Rulers are made of elementary particles known as fermions, which cannot move on their own even if smashed into sub-particles, and whose flight requires foreign energy. The faster it travels, the more energy it needs, an infinite amount of energy to reach the speed of light — this is a theory of corpuscles that essentially negates the law of mass-energy conversion. Matter is matter, and its carrier is fermion; Energy is energy, and its carrier is the boson, which is so different in nature that it cannot be converted even at the speed of light.

My theory of corpuscle is the theory of photons, which is itself a combination of mass and energy, and the speed of light is the dividing line of its two states.

Here, I highlight a previous point: Do not regarding particles and energy as two isolated things that are constantly changing into each other. **Specifically, do not think of particles as hard grains like sand and energy as an elusive gas; they are merely photons at rest or in flight.** Although particles have shells (boundaries), they are not composed of foreign hard stuffs, but are formed in their own swirling motion, "all boundaries are ultimately dynamic." (Bertalanffy, 1969, p215) As particles move faster, the binding force of the shell becomes weaker, and finally disappears, the photons are completely free, and the matter is thus transformed into pure energy. This is a great secret of the cosmos.

Energy-Mass Conversion: Changing Mass with Speed

The unlike is joined together, and from differences results the most beautiful harmony.
All things come into being by conflict of opposites.

Heraclitus, *Fragments*

Stephen Hawking believes that there are 10^{80} particles in the universe. Where does so much matter come from? The source could only be energy. After discovering dark energy, scientists generally believe that energy accounts for more than 90% of the total energy-mass and that the rest is matter, which is the basis for energy, to produce matter.

How can energy be converted into matter? Einstein founded the theory of mass–energy transfer and did not answer this question. His mass–energy formula is irreversible. Hawking mentioned the answer of quantum theory in his book: particles can be created out of energy in the form of particle–antiparticle pairs. (2016, p.146) He continued. "Then, as the universe expanded and cooled, the antiquarks would annihilate with the quarks, but since there would be more quarks than antiquarks, a small excess of quarks would remain. It is these that make up the matter we see today and out of which we ourselves are made." (p.89).

I think that free quarks are not matter yet. Matter is the state of particle aggregation and confinement, and its smallest unit is atoms. This produces two forms of energy-mass conversion.

First, a concentrated, large-scale format. All particles are originally photons, and the fundamental form of photons is the field of light. The light field is always in motion, which means it is definitely uneven. Like inside a collider, starlight from different directions collides violently in a star-studded space, will form a large amount of interstellar matte. I also ascribe it to 'light barrier' mass products. (ref. 'Cosmic Field and Material Fields')

Second, the scattered moments in which matter is produced by the interaction of light and matter. In the solar system, for example, the Sun is constantly raising the speed of molecules in the planet's atmosphere and crust in the form of radiant light, and promotes the movement of polymerization, including particle polymerization and photosynthesis. In this way, solar energy is converted into planetary energy and matter. Scientists also told us that the solar wind is mainly composed of particles such as protons and electrons, which are products of energy-mass conversion. This continuous stream of particles is also the inexhaustible source of new material for planets.

From the description in this section and the previous section, it can be seen that **energy-mass conversion & mass-energy conversion are very common universal motions that are going on all the time.** Particles are not the product of a certain quantity formed at the time of the big bang, but the product of the light field. The key point is that we should regard photons as energy-mass continuum; the speed of light is their energetic state, and particle is their material state. They are two consecutive states of the same photon.

I have always been proud of this unique idea. But recently I read *The Principia*: Introduction To The American Edition, and found that Newton had this conjecture, although his basic theory is different from mine, which dispel my pride.

"He regarded light as consisting of small material particles emitted from shining substances. He thought that these particles could be re-combined into solid matter, so that **'gross bodies and light were convertible into one another;'** that the particles of light and the particles of solid bodies acted mutually upon each other; those of light agitating and heating those of solid bodies, and the latter attracting and repelling the former. Newton was the first to suggest the idea of the Polarization of light." (*The Principia*, p.15)

I believe that most people still ignore or don't understand Newton's this unique idea, which reminds me of Plato's assertion that **if a person lacks the highest knowledge, he can't learn other knowledge well and can't benefit from them.** (*The Republic* 505a)

Newton actually used here the familiar causal loop, or logical reasoning. Like the chicken and the egg⌢the egg and the chicken; special plants produce special seeds ⌢special seeds grow into special plants; the decomposition of hydrogen and oxygen from water⌢the combination of hydrogen and oxygen into water; water is agitating to form steam, which is cooling and returned to water, etc. All are certain.

Newton didn't seem to have enough confidence in it, so he didn't include this topic in his *Optics* or other works.

Cracking the Myth of the Double-Slit Experiment

More than two hundred years ago, in what is known as the double-slit experiment, a British physician called Thomas Young proved that light, not particles, was a wave when he performed an optical experiment, showing that a beam of light passing through two slits would become an interference fringe. Some people repeated the experiment by emitting a single photon, and the results remained the same. This phenomenon has puzzled scientists up to today, especially the latter case. The photon, thought to be an indivisible sub-particle akin to the electron, now exhibits powerful body-splitting and self-interfering behaviour — **"each electron, therefore, must be passing through both slits at the same time!"** (Hawking, 2016, p.68)

Instead, scientists experimented with single neutrons, atoms, and many molecular types, even large molecules such as buckminsterfullerene — made of 60 carbon atoms — will behave in this way. In *Fields of Color* (p.67-68), Rodney A. Brooks tells us that this phenomenon puzzled Einstein for the rest of his life.

"We are forced to conclude that... a single photon is responsible for the ability of the two beams to interfere, as well as for the absorption of light from one of the beams. It is evident that Maxwell's theory cannot account for this complex of properties of the photon. It does not provide us with any means to understand the atomistic character of the absorbed energy of radiation... [Yet] the interpretation of the photon as a point like structure does not admit of an explanation for the interference phenomena which are produced if both parts of the beam interact. — A. Einstein (1950 speech to International Congress of Surgeons, quoted in Physics Today, June 2005, p. 47)

He spent the last half of his life searching for a unified field theory that would explain (among other things) the peculiar behavior of light quanta. Shortly before his death he wrote: **'All these fifty years of pondering have not brought me any closer to answering the question, what are light quanta.'** A. Einstein (P1982, p. 382)"

Philip Ball, a British science writer, put forward two oddities of this experiment in one of his articles:

"What's odd is that the interference pattern remains — accumulating over many particle impacts — even if particles go through the slits one at a time. The particles seem to interfere with themselves. Odder, the pattern vanishes if we use a detector to measure which slit the particle goes through; it's truly particle-like, with no more waviness. Oddest of all, that remains true if we delay the measurement until after the particle has traversed the slits (but before it hits the screen). And if we make the measurement but then delete the result without looking at it, interference returns." (*Nature*, 2018)

The answer provided by quantum scientists, quantum collapse, is not persuasive. Applying the LFT and the new interpretation of the wave– particle duality of light, these weird phenomena were instantly and easily solved.

The light field is full of space, fill all the gaps and corners, and the photon is instantaneously converted into part of the light field through mass-energy conversion after it is launched. **Its energy, in the form of waves, propagates and intensifies in the light field, forming a series of resonant waves and causing interference when they encounter a double slit. At the end, when these light waves fall on the screen, interacts with the screen, energy-mass conversion occurs, light waves are trapped and become photons, and the screen displays the result.**

That is to say, it is not a moving image of one photon or many photons that is presented to us, but an image in which the light field interacts with the screen after being disturbed by one or more photons, or images of photons emitted dragging the light field to fly. **In short, energy, even from a single corpuscle, needs to be transmitted as waves through the light field, not as a single particle flying itself.** It is as if the energy of a raindrop returning to the sea is transmitted not by itself but by the waves of the sea water.

Remember, a particle is a particle, not a wave; There are no particles in the field, only waves. Just like a raindrop is a dot, not a plane; There are no water drops in the ocean, only waves.

CFT determines that scattered corpuscles cannot form fields and waves — this is QFT (Quantum Field Theory). "In QFT the photon is a spread-out field" (Brooks, 2016, p.68) Therefore, QFT believes that a single corpuscle can be passing through both slits at the same time.

So, that explains the first oddity. The second oddity is easy to understand when we apply CFFT (cosmic force field theory). The detector is similar to a monitor or radar and has **a scanning function, so that the radiation will eliminate the light waves.** Even if it has no scanning function, and just simply receives information, it itself also has radiation, which will produce the same effect as soon as it is turned on. Going back to the previous example of throwing a pebble into the water, if it's a rainy day (equivalent to detector radiation) and you throw the pebble (equivalent to the single particle being emitted) into the water, you'll only see a bigger drop, not spreading waves, which is lost in mutual interference or becomes blurred.

NOTE. with reference to the data provided by Frank Wilczek, I set the ratio of the EEEF at 1:30,000, or more. Thus, a photon transmitted through the field would appear on the screen as 30,000 photons, and a set of interference fringes would be understandable—I don't appreciate this solution, it is still a MBT. Light field is my first choice. Without the light field, the behaviors of light as well as particles are weird and incomprehensible.

History of the Vortex Theory

From Diogenes Laërtius' famous book, *The Lives and Opinions of Eminent Philosophers*, I find that Leucippus and Democritus are the founders not only of atomism, but also of vortex theory. In their view, the basic form of motion of atoms is vortex movement, which forms the inner shell and many outer shells of things. All stars are formed by the contraction of the shell, so they are spherical systems. The following is the original text:

"He [Leucippus] says that the universe is infinite, as I have already mentioned; that of it, one part is a plenum, and the other a vacuum. He also says that the elements, and the worlds which are derived from them, are infinite, and are dissolved again into them; and that the worlds are produced in this manner: That many bodies, of various kinds and shapes, are borne by amputation from the infinite, into a vast vacuum; and then, **they being collected together, produce one vortex;** according to which they, dashing against one another, and whirling about in every direction, are separated in such a way that like attaches itself to like. But as they are all of equal weight, when by reason of their number they are no longer able to whirl about, the thin ones depart into the outer vacuum, as if they bounded through, and the others remain behind, and becoming entangled with one another, run together, and produce a sort of spherical shaped figure. **This subsists as a kind of membrane; containing within itself bodies of every kind;** and as these are whirled about so as to revolve according to the resistance of the centre, the circumambient membrane becomes thin, since bodies are without ceasing, **uniting according to the impulse given by the vortex;** and in this way the Earth is produced, since these bodies which have once been brought to the centre remain there.

On the other side, there is produced another enveloping membrane, which increases incessantly by the accretion of exterior bodies; and which, as it is itself animated by a circular movement, drags with it, and adds to itself, everything it meets with; some of these bodies thus enveloped re-unite again and form compounds, which are at first moist and clayey, but soon becoming dry, and being drawn on in the universal movement of the circular vortex, they catch fire, and constitute the substance of the stars. The orbit of the sun is the most distant one; that of the moon is the nearest to the earth; and between the two are the orbits of the other stars." (Laërtius, 2006, p.32-34)

"Democritus, now his principal doctrines were these. That atoms and the vacuum were the beginning of the universe; and that everything else existed only in opinion. That the worlds were infinite, created, and perishable.

But that nothing was created out of nothing, and that nothing was destroyed so as to become nothing. **That the atoms were infinite both in magnitude and number, and were borne about through the universe in endless revolutions.** And that thus they produced all the combinations that exist; fire, water, air, and earth; for that all these things are only combinations of certain atoms; which combinations are incapable of being affected by external circumstances, and are unchangeable by reason of their solidity. Also, that the sun and the moon are formed by such revolutions and round bodies; and in like manner the soul is produced; and that the soul and the mind are identical: that we see by the falling of visions across our sight; and that everything that happens, happens of necessity. Motion, being the cause of the production of everything, which he calls necessity." (p.44-46)

We are used to thinking of Democritus and Leucippus as cofounders of atomism & vortex theory, which is not appropriate because they were not contemporaries.

Leucippus, forty years older than Democritus, must be the originator of atomism and vortex theory, with the focus on vortex theory, whereas Democritus could only be considered a supporter and promoter, with an focus on atomism.

According to them, atoms are the most primitive and fundamental substances in the universe. When they flood into a vacuum, they form a vortex and a shell(membrane), and evolve into elements and everything within. Their theory might be called the atomic vortex theory, and it was crude but significant.

1. Vortex theory was first proposed.
   Greek philosophers put forward a variety of original theories, such as Thales' water element theory, Heraclitus' fire element theory, Pythagoras' number theory of everything, and so on, contemporary astronomical observations strongly support the vortex theory, Leucippus is the founder of atomism and vortex theory.
2. Created the first vortex theory that came close to reality.
   The vortex described by them is rolled inward, and under the constraint of membrane, heavy atoms gather in the center and light atoms are scattered around; In other words, the outer density is lower than the inner density. In addition, the rotation and combination of atoms depend on the power of vortex itself ("the impulse given by the vortex", and "Motion, being the cause of the production of everything, which he calls necessity"), without external force or ether as medium. This description is very close to the observations and experiments of atomic motion by contemporary scientists, and also close to the CFVT of this book, which is more reasonable than the vortex theories of Descartes and Newton.
3. Second evidence for the superiority of speculative philosophy over experimental philosophy.

The first evidence is Euclidean geometry. "Once these axioms were selected, however, the theorems were deduced by the action of the mind alone. Each and every one of the hundreds of theorems contained in the Elements could have been deduced by a Euclid sitting with blindfolded eyes in an ivory tower." (Kline, 1953, p.81)

At that time, there are no microscopes and telescopes yet, which are the products of science and technology 2200 years later. Their vortex theory is neither the result of observation nor the result of scientific experiment, but the result of **philosophical speculation.**

This fact confirms a phrase in Hawking's book: "The Aristotelian tradition also held that **one could work out all the laws that govern the universe by pure thought:** it was not necessary to check by observation." (2016, p.17) And in Descartes': "we must note that while our experiences of things are often deceptive, the deduction or pure inference of one thing from another can never be performed wrongly by an intellect which is in the least degree rational though we may fail to make the inference if we do not see it." (Descartes, 1985, p.12)

This is how holism works, it reminds us of Einstein's famous saying, **"Imagination is more important than knowledge."** Although it has many defects, as described in the same page of Hawking's book, it can not tell 'whether bodies of different weight did in fact fall at different speeds', but **it can discover the truth with ease and relative assurance.** Most truths and laws in history, including atomism and relativity, were discovered by philosophers and scientists in this way.

Influenced by Democritus and Newton, most people attach importance to atomism and neglect the vortex theory, which is the habit of MBT. Only a few philosophers explicitly support the vortex theory.

Giordano Bruno, the Italian Dominican friar, philosopher, mathematician, poet, cosmological theorist, was an early supporter of the vortex theory, then Descartes, Kant, and Laplace. Among them, Descartes did the most thorough and most developed research on the theory of vortex. He called the motion of vortex 'the eternal movement of the physical universe', which was the highest evaluation of this motion in his time. Following the basic premises of Descartes, Christiaan Huygens, the Dutch physicist, mathematician, astronomer and inventor, between 1669 and 1690 designed a much more exact vortex model. This model was the first theory of gravitation which was worked out mathematically.

Descartes' vortex theory had been popular in Europe for decades. But he was unlucky. By the early 18th century, his theory had been superseded by Newton's gravity theory. Newton, for his laws of motions, also established a fluid vortex theory in *The Principia*. However, this vortex theory has received little attention from successors, who prefer to use his gravitational theory to explain the vortex motion and the origin of the universe instead of the fluid vortex theory. Their vortex theories will discuss in detail later.

"The introduction of the pure and lofty doctrines of *the Principia* was perseveringly resisted...So that, the saying of Voltaire is probably true, that though Newton survived the publication of his great work more than forty years, yet, at the time of his death, he had not above twenty followers out of England. But in England, the reception of our author's philosophy was rapid and triumphant." (*The Principia*, p.23)

After gaining strong support from Kant, Newton's theory was widely popular in Europe. The full name of Kant's cosmological work is called *Universal Natural History and Theory of the Heavens* (German: *Allgemeine Naturgeschichte und Theorie des Himmels*), or *An Attempt to Account for the Constitutional and Mechanical Origin of the Universe Upon Newtonian Principles*.

Related to the vortex theory is the nebular hypothesis, which was first proposed in 1734 by Swedish scientist Emanuel Swedenborg and later expanded upon by

Immanuel Kant in 1755. A similar theory was independently formulated by the Frenchman Pierre-Simon Laplace in 1796. In 1944, the German physicist and philosopher Carl Friedrich von Weizsäcker developed a theory on the formation of the Solar System, which went back to the Cartesian model by involving a pattern of turbulence-induced eddies in a Laplacian nebular disc. It was much criticized, as turbulence is a phenomenon associated with disorder and would not spontaneously produce the highly-ordered structure required by the hypothesis. It also does not provide a solution to the angular momentum problem or explain lunar formation and other very basic characteristics of the Solar System.

More than 30 years ago, I bought Kant's *Universal Natural History and Theory of the Heavens* (Chinese version) and started thinking about the motions of the heavenly bodies. With the advance of observing sky in the world, my theory gradually formed, and the result was *The Force Cosmology* in 2017. The book was aborted because of the false ISBN used by the publisher. After that, I contacted other publishers in China, but none of them was interested in the book, so I began to turn to foreign publishers. *An Outline of Force Cosmology*(2019) and this book are published smoothly. **I'm trying to guide cosmology back to Leucippus at a new height.**

Absolute Truth: Cosmic Force Vortex Theory (CFVT)

"Absolute truth is something that is true at all times and in all places. It is something that is always true no matter what the circumstances. It is a fact that cannot be changed. Absolute truths are discovered, not invented."
(Wikipedia Encyclopedia)

Ninety years ago, Edwin Hubble created the Hubble tuning fork. At the same time, he provided a set of astronomical observations alleging that **in all visible galaxies, spiral galaxies account for 50% of the total; barred spiral galaxies, 30%; elliptical galaxies, 17%; and irregular galaxies, 3%.** This idea has been universally accepted by astronomers. Only a small improvement needs to be made. (Prof. Richard Pogge, Lecture 27: Spirals & Ellipticals & Irregulars. http://www.astronomy.ohio-state.edu/~pogge/Ast162/ Unit4/types.html)

We know that barred spiral galaxies are vortex bodies with two cores revolving around each other, which are not essentially different from single-core vortex bodies. The irregular galaxy is the early stage of the spiral galaxy, and the elliptical galaxy is its late stage.

One conclusion: **All galaxies are spiral galaxies.** All astrophysical systems, lower than the galaxy and higher than it are in, or have been in, vortex motion. The universe is a big entity where bigger vortexes envelop smaller vortexes — vortexes everywhere.

In recent years, quantum scientists have also observed by microscopes an atom's small, invisible tornado-like vortex, with a number of infinitely small energy vortexes, called quarks and photons. "Figure 6.1 Pictures of the interior of a proton. a. A proton moving at nearly the speed of light appears flattened in the direction of motion, according to the theory of relativity. b... Inside are quarks and gluons, also moving at nearly the speed of light. They share the total energy of the proton, and the sizes of the arrows indicate their relative shares." (Wilczek, 2008, p.42) "But this is a creative destruction that allows new imaginative constructions. For example, it enables us to reconcile two seemingly contradictory ideas about what protons are. On the one hand, the interior of a proton is a dynamic place, with things changing and moving around. On the other hand, all protons everywhere and every when behave in exactly the same way...It is like a smoothly flowing river, which always looks the same even though every drop of it is in flux." (Ibid, p.43) This statement is also suitable for vortex motion. Proton has the strongest shell, and all of its internal motion is confined by the shell and can only be vortex. In other words, **not gluons or gravitons, but vortex motion, the one-way wind-up motion, makes proton unbreakable.**

A cosmological truth that comes out of this: vortex motion, from particle spin to celestial vortex curl, is the most basic and universal movement of the universe. Vortex force is the greatest force of the universe. The universe is a maelstrom.

We also know, from observation and common sense, that all vortexes move in a centripetal constriction, hence another conclusion that the universe is contracting everywhere, and it is impossible for the whole to expand continuously.

Vortex force comes from the aggregate energy of particles and increases with the accumulation of particles. Therefore, I call it cosmic force vortex theory (CFVT) so as to distinguish it from several previous vortex theories, as they all require an external force or God's divine impetus.

The following examples will help deepen your understanding. If space has a pile of sand with the same mass as the Sun, even if you put a crust on it to prevent losing any sand, no matter how long it takes, it won't automatically rotate and form a bright sun. According to Newton's laws, it has a gravitational force corresponding to its mass that should evolve into a sun. Only when each grain of sand in the sand pile gains speed — that is energy — and is under the constraint of the shell (crust), can the pile of sand, without light, become a luminous body. The aggregation of particles is this effect, and its energy is not foreign but intrinsic, because all particles come from photons, and light is energy.

Sand is also made of particles, but it is unlike nebulae. While the sand particles are trapped in their atoms and cannot move freely, the particles in the nebula are completely free. Even if this does not account for all of them, there must be a considerable number of free particles. If the crust envisioned above keeps rotating and contracting, breaking down the sand pile's molecular and atomic structures and freeing the sand particles inside, the dull sand will also become a luminous body, which is a crude mode of CFVT.

Truth comes from faith, and faith is based on facts. Fluxionary facts produce relative truth, and eternal facts produce absolute truth. CFT and CFVT are founded on eternal and infinite facts without exception, so they are absolute truths.

And **absolute truth is a great challenge to human intelligence. Readers are entailed to either support or refute it**—Just ask yourself and answer, not necessarily tell anyone, if you don't like to.

The Direction of the Hubble Tuning Fork Is Wrong

Edwin Hubble was the first person to classify galaxies morphologically, and astronomers used, and continue to use, his system called the Hubble tuning fork. Hubble divided the galaxies into two general categories: elliptical galaxies and spiral galaxies. They meet at the lenticular galaxy and are subdivided below the two.

Spiral galaxies include barred spiral galaxies, which are characterized by one or more arms winding in toward their bright centre, while elliptical galaxies have no such arms. A small number of irregular galaxies were thought by Hubble to be the result of collisions between the first two types of galaxies or of their gravitational pull.

Hubble believes that galaxies evolve from left to right in the diagram of the tuning fork. The spiral arm grows in the process of evolution from short to long, from tight to loose. So, he called elliptical galaxies 'early' galaxies, and spiral galaxies 'late' galaxies. This later results in the universe expansion theory.

Astronomers now know that Hubble was wrong, even though they still believe in the big bang theory. This is because spiral galaxies rotate quickly, while elliptical galaxies do not. There is no way that an elliptical galaxy could spontaneously begin rotating, so there is no way an elliptical galaxy could turn into a spiral galaxy. But scientists still use Hubble's terminology today; elliptical galaxies are still referred to as early galaxies, and spirals as late galaxies. The theory of cosmic expansion is also popular and has since been developed.

However, it is not uncommon for a high-speed spinning red or blue giant to turn into a spiral galaxy or a cluster after a series of explosions (supernova explosions). These giant celestial bodies usually born in this way. But obviously, it is not the subject of Hubble's diagram, which is based on all galaxies that astronomers have observed at that time.

The most important reason for the popularity of cosmic expansion comes from the understanding of redshift and quasar, which needs to be changed in two ways. First, the Doppler effect applies to acoustics, not optics, because the concepts of sound and light have nothing in common. The former is a motion of matter, the latter is a motion of energy, no physical law can be applied to both. In addition, according to Einstein's proposition that the speed of light has nothing to do with the light source's motion, resulting in no red or violet shift from it. Actually, the redshift is mainly caused by quasars, which is a phenomenon of stars being torn apart violently, resulting in huge redshifts. This is also the reason why astronomers have observed a large amount of redshift but few violetshift.

Tracing back to the source, we will not be difficult to find that **the direction of Hubble tuning fork originated from Newton.** In *The Principia*, Newton conceived **"a vortex, and that motion will by degrees be propagated onward in infinitum."** (*The Principia*, 380). The more basic reason is that "the centripetal force. And the same thing is to be understood of all bodies, revolved in any orbits. They all endeavour to recede from the centres of their orbits." (Ibid.,3) This will be discussed in detail later.

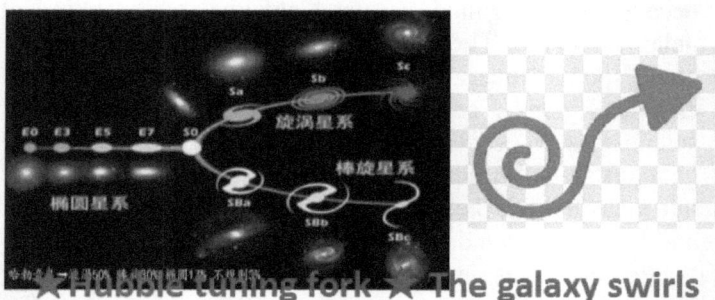

Hubble tuning fork ★ The galaxy swirls from tight to loose

Design by Solatle Lu

Constant Speed of Light ➤ Doppler Effect Becomes Ineffective

"The American astronomer Edwin Powell Hubble reported in 1929 that the distant galaxies were receding from the Milky Way system, in which Earth is located, and that their redshifts increase proportionally with their increasing distance. This generalization became the basis for what is called Hubble's law, which correlates the recessional velocity of a galaxy with its distance from Earth. That is to say, **the greater the redshift manifested by light emanating from such an object, the greater the distance of the object and the larger its recessional velocity** (see also Hubble's constant). This law of redshifts has been confirmed by subsequent research and provides the cornerstone of modern relativistic cosmological theories that postulate that the universe is expanding.

Since the early 1960s astronomers have discovered cosmic objects known as quasars that exhibit larger redshifts than any of the remotest galaxies previously observed. The extremely large redshifts of various quasars suggest that they are moving away from Earth at tremendous velocities (i.e., approximately 90 percent the speed of light) and thereby constitute some of the most distant objects in the universe."

(Encyclopedia Britannica)

In a nutshell. By observing hundreds of galaxies and determining their redshifts, Hubble created Hubble's law, and inferred that they were receding at high speed from us and that the universe was expanding. Its theoretical basis is the Doppler Effect, abbreviated as DE.

"In 1842 Austrian physicist Christian Doppler established that the apparent frequency of sound waves from an approaching source is greater than the frequency emitted by the source and that the apparent frequency of a receding source is lower. The

Doppler effect, which is easily noticed with approaching or receding police sirens, also applies to light waves. The light from an approaching source is shifted up in frequency, or violetshifted, while light from a receding source is shifted down in frequency, or redshifted. The frequency shift depends on the velocity of the source relative to the observer; for velocities much less than the speed of light, the shift is proportional to the velocity."

(Encyclopedia Britannica)

That is to say, the DE is applicable to the whistle and light of the patrol wagon, the principle of sound DE and light DE is the same. However, Einstein's theory of the constant speed of light denies the light DE.

"No matter how you measure it, the speed of light is always the same.

Einstein's crucial breakthrough about the nature of light, made in 1905, can be summed up in a deceptively simple statement: The speed of light is constant. So what does this sentence really mean?

Surprisingly, the answer has nothing to do with the actual speed of light, which is 300,000 kilometers per second (186,000 miles per second) through the 'vacuum' of empty space. Instead, Einstein had an unexpected—and paradoxical—insight: that light from a moving source has the same velocity as light from a stationary source. For example, beams of light from a lighthouse, from a speeding car's headlights and from the lights on a supersonic jet all travel at a constant rate as measured by all observers—despite differences in how fast the sources of these beams move.

Moving Light
Light from a moving source also travels at 300,000 km/sec (186,000 miles/sec).

Say that Einstein's bike travels at 10% the speed of light (30,000 km/sec): the speed of light from Einstein's headlight does NOT equal 330,000 km/sec.

The speed of light is constant and does not depend on the speed of the light source." (American Museum of Natural History https://www.amnh.org/exhibitions/einstein/)

The sirens of the police car has a strong DE, but it headlights have no DE at all, The reason is that optical DE and acoustic DE are two completely different physical concepts.

Light is cosmic energy, moving at a constant ultimate velocity, self-propelled. Sound is the motion of the air squeezed by the sound source, the motion of molecular waves, the motion of matter, depending on external forces, **the two kinds of motion are completely different in nature. The former is the motion of energy field and the latter is the motion of material field, they obey different laws of motion.** Even in comparison with the speed of motion, the speed of light is 300000000m/s, and the speed of sound is about 340m/s, equivalent to one millionth of the speed of light. The two are respectively like a train speeding along a track and a tortoise crawling on the ground. In the eyes of a passenger on a train, the tortoise is motionless or receding. It makes no sense to apply to the former the laws of motion that apply to the latter.

Similar to acoustic DE, fluid DE. When a ship sails on the water surface, it will generate water waves. The wave in front must have a shorter wavelength and a higher frequency than the wave behind, that is, the wave in front is high and strong, while the wave behind is flat and weak. It's also molecular motion, medium motion. The motion of water waves is much slower than that of sound waves and is also negligible compared with light waves.

From LFT, we can regard the violetshift as EEEF, and redshift as DEEF (decaying energy effect of field). Then, we can find, neither sound wave nor water wave can derive any effect from light field, because they do not conform to the 'synchronization law'. Besides, they are material fields, which conform to the 'irrelevance theorem'.

Light Waves Get Longer as They Travel and All Starlight Is Redshifting

When you enjoy a movie in the cinema, do you think that the area of the projector lens is only about 600 square millimeters, or 6 square centimeters, and the light source shining on it is also such an area, but it creates a wide moving image for the audience to enjoy — if you measure them with precision instruments, you will be surprised to find that the light on the screen is always redshifting!

I haven't made this measurement, but I am sure of this result, because the wavelength and frequency of light waves on the screen must be different from those of light sources. This change must be that the wavelength becomes longer and the frequency decreases, but not the opposite, that is, only redshifts will occur instead of violetshifts.

The reason is simple but unarguable: the tiny spot of light needs to be spread out over a screen thousands of times larger than its original area, and the only way to do that is to expand the wavelength of the light and reduce its frequency. **It is based on the principle that the speed of light (c) is constant, and c =λf. If its wavelength (λ) increases, the frequency (f) decreases accordingly, vice versa.**

We can also prove this intuitively. We can't watch this light source for a few minutes, but we can enjoy the image on the screen for hours, just because the former light is much more intense than the latter, with a shorter wavelength and higher frequency. There is also the second law of thermodynamics, which tells us that heat must be cooled in spacetime, that light must be dimmed rather than brighter as it travels, and that it dims faster, with violet light suddenly turning red.

This phenomenon, in turn, is related to the density of the light field; the light field of the light source is nearly ten thousand times denser than that of the screen. In other words,

the photons emitted from the light source are scattered ten thousand times over the space, so that the brightness of the light drops, and the color turning red, is inevitable. A convex lens concentrates light, making the light wave shorter and increasing its frequency. A concave lens astigmates, making the light wave longer and decreasing its frequency, which is also a result of the change in field density.

Contrary to what some quantum scientists believe, when they assume that photons are conscious, this change does not occur on purpose, it occurs naturally as photons pass through mechanical settings. This machine was designed and built but by engineers who unwittingly followed the principle of light ball motion.

The theory of light ball is a theory founded in this book. Its principle is unfamiliar to most people, but it is often used unconsciously. It has something to do with Hubble's law.

All natural light sources, especially starlight, emit light from the inside to the outside in a spherical circle, and the outer circle is always larger than the inner circle. Inevitably, the light wave will be stretched continuously as the radius of the light sphere increases during transmission, and the frequency decreases correspondingly, showing a redshift, and **the redshift value of stars is directly proportional to its spherical radius.** It goes without saying that if we know the redshift value of a star, we can calculate its distance from us.

Sunlight, for example, must have redshifted when it reached Earth 8.3 minutes later, crossing a distance of nearly 150 million kilometres. Astronomers set this distance at 1 astronomical unit (AU), and if we set its redshift to 1 accordingly, Neptune is about 30 AU from the sun, and the redshift of the sunlight above it is 30. Conversely, if we measured this redshift as 30, we would also know its distance from the Sun. This method applies to allstarlight.

Using this calculation, we can get the same result as Hubble: the farther away from the solar system, the greater the redshift value of galaxies. But this redshift has nothing to do with the Doppler effect. Stars aren't receding from Earth, and the universe isn't expanding — just the ball of light is expanding.

The exception is quasars, whose large redshifts are caused not only by distance but more importantly by being torn apart. The brightness of light varies so dramatically that astronomers are often able to measure faster-than-light redshifts.

That is to say, the brightness of stars is another factor of redshift, which is positively correlated. All stars have redshift, however, in the same space, the brighter and bluer stars have relatively higher redshift, and the red and dim stars have relatively lower redshift. **Most quasars are the brightest, most dramatic objects in the sky, so they have the highest redshift.**

In addition, there is a third factor: **EEEF. This force is one hundred times stronger than that of the most powerful accelerator in the world, which eliminates most of the redshift and even reverses it.** Astronomers occasionally observe the phenomenon of starlight violetshifts, which should be caused by local super EEEF. Without EEEF, we would not see most of the starlights.

TIP. Why do light waves get longer and less frequent in flight?

The light ball expands at twice the speed of light. Each light source on earth, strong or weak, a nuclear bomb explosion or a candlelight, must produce a ball of light 3×10^5 km in radius every second. Its volume (V1) is calculated as follow.

The formula for calculating the volume of a sphere:
$V0 = (4/3)\pi R^3 = 4.1889 R^3$, then
$V1 = 4.1889 \times (3 \times 10^5 \text{ km})^3$
$= 113.1 \times 10^{15} \text{ km}^3 = 1.131 \times 10^{17} \text{ km}^3$
Scientists calculate that the volume of the earth is
 $1.083 \times 10^{12} \text{ km}^3$

As a result of comparison, the number of the earth that can be accommodated in the light ball formed by any light source in 1 second is 1.044×10^5, viz. 104,400.

After another 0.3 seconds, this ball of light expands to 780,000 km in diameter, enough for the earth-moon system (760,000 km in diameter) to operate inside.

The Differences Between the Ground Vortex and the Sky Vortex

There are two main types of vortexes on earth. One is a gas vortex, typically a tornado. The other is a liquid vortex, a water swirl. These two vortexes belong to the polymerization movement of molecules with a weaker cohesive force. Their main driving force is **an external force,** such as solar energy and potential energy, which will disappear with the exhaustion of the energy.

The vortexes in sky are mainly the vortexes of stars and galaxies, and they belong to the aggregated movement of particles. **Their force is provided by particles themselves, and the energy never decays**, let alone dies. On the contrary, these vortexes are strengthened by the growth of the vortex and the universal contraction effect. They are the original source of everything, and the force of everything.

Until now, only galaxy vortexes have been observed. A stellar vortex consists of nonluminous matter, and its mass is much smaller than that of a galaxy vortex, which has not yet been noticed or observed by astronomers. In my opinion, all red giants are observed as stellar vortexes that have gathered their arms (Compared to galaxy vortexes, their arms are so short and fuzzy that we can hardly see them), our early sun was a red giant, and Betelgeuse is a red giant star that newly evolved from a stellar vortex. In fact, there are far more stellar vortexes in the sky than galaxy vortexes, but no one has seen them yet. The first reason here is that whoever wants to see them must first believe in the CFVT.

Earth's vortex, such as a tornado, has entrance and exit, and the material that swirls toward the centre will be lost at last. There are many obstacles on its way, it will disappear sooner or later. **The vortex motion in sky has no resistance, so it will never disappear, only change its shape. It is also a closed movement. The material has only the entrance**

and no exit. The material being rolled toward the centre has no way out, it can only contribute to the formation of the core — the antecedent of heavenly bodies. Therefore, as the core is growing, density and pressure are rising, meaning the nature of the two kinds of motion is completely different. The ground vortex is the accidental form of the movement of existing objects, while the space vortex is the process of the evolution of the original material itself.

For this characteristic, space vortexes can have a complete evolution process, galaxy vortexes being the most obvious one. From the early irregular shape, to the core and spiral arm of the whirlpool, to the later ellipse, this is the reverse sequence of the Hubble tuning fork.

High-level vortex systems evolve simultaneously. Star vortexes combine to form galaxy vortexes, galaxy vortexes combine to form super-galaxy vortexes, and so on, and finally the universe evolves into a huge vortex — a maelstrom.

In *Relativity* (2006), Einstein suggested that the Milky Way and the Andromeda belong to a local galaxy group, which moves at a speed of 600 kilometres per second to the local supercluster of Virgo. In this supercluster, Ophiuchus and Centauri fall to another larger galaxy group, which astronomers call the 'grand gravitational source'. These clusters, along with superclusters, form a range of matric and filamentous structures that are hundreds of millions of light-years away, resembling cellular tissue in living organisms. I call them 'two-level vortexes' above the Milky Way.

The Galactic Density Wave Theory Lacks Persuasion

Most spiral galaxies have one or more spiral arms, but no astronomer believes that spiral arms are real. Instead, density wave theory is sought after. This theory, which relies on hydrodynamics, holds that galaxies rotate to produce density waves, which stars encounter as they rotate around the centre of a galaxy, then they slow down a bit and assemble into **a spiral arm pattern.** Eventually, they will cross the arm, return to their original orbits and speeds, and continue their inherent journey, never getting sucked into the centre.

Compared with the spiral arms, rivers on the ground look small, but they all return to the sea at last, and the sand and mud along the way are wrapped in them, unable to escape, eventually flow into the sea, which is consistent in this respect. According to the density wave theory, it is equivalent to saying that the river water and the things trapped in it are not the inherent things of the river, but the things flowing horizontally in many rivers that will never converge with the sea — what an incredible picture!

The Milky Way is a big vortex. Astronomical observations show that the orbital motion of stars in the system is accelerated step by step from outside to inside, and the density of celestial bodies is also intensified step by step from outside to inside, showing that the density and velocity are synchronized, which completely violates the density wave theory (density slows down the stars), but completely conforms to the law of vortex motion.

Density wave theory is particularly difficult when you come to explain the phenomenon of **the single-arm whirlpool,** which is particularly like the spring of a mechanical clock. How did it come into being? How could density waves bring almost all the stars together on a single arm, and what forces would make the stars on the arm disengage in the future and return to their original orbits?

There is no doubt that the single spiral arm will wind all the way to the centre, forming a star, like a tightened clockwork spring. The same is true of more spiral arms.

Here is a fundamental difference between me and astronomers. They clearly see the swirling galaxies all over the sky, but no one believes in the centripetal motion of the stars on the spiral arms. Whereas I have never seen a galaxy spiral with my own eyes, and my whole knowledge of astronomy came from astronomers and philosophers, but **I firmly believe that all stars on all spiral arms are rotating and precessing to their centres without exception, until finally they all fall into the central black hole, forming new celestial bodies.**

Hotline public picture

Spiral Shell Theory (SST)

"BOUNDARIES. Any system as an entity which can be investigated in its own right must have boundaries, either spatial or dynamic. Strictly speaking, spatial boundaries exist only in naive observation, and **all boundaries are ultimately dynamic.** One cannot exactly draw the boundaries of an atom (with valences sticking out, as it were, to attract other atoms), of a stone (an aggregate of molecules and atoms which mostly consist of empty space, with particles in planetary distances), or of an organism (continually exchanging matter with environment)." (Bertalanffy, 1969, p215)

This theory derives **from Leucippus' enveloping membrane hypothesis, which is the boundary of all kinds of vortexes and an indispensable condition for the movement of vortexes.** It forms in two ways. One way is the inner edge of matter that the vortex hasn't connected to or disconnected with, such as the atmosphere that surrounds the tornado, and the nebula that envelops the stellar vortex. They diffuse in a vast space, and there are always places where vortexes cannot reach. The second is the result of its own movement. One feature of vortex motion is that the rotation speed of each spiral inner-circle is higher than that of the adjacent outer circle, and the whole is a gradually contracting low- pressure body. The central black hole has the lowest pressure, and each outer circle thus becomes the shell of all inner circles and the black hole. From the appearance of some shells, some people also call them rings or loops.

Stephen Hawking has a book called *The Universe in a Nutshell*. Chapter seven, 'The New World of Membranes', can be regarded as a mathematical description of Leucippus' theory. The high dimension proposed in his book can be understood as the overlapping membranes between celestial bodies. But I'm not sure he agreed with spiral shell theory.

The role of the shell on the object is dual, both restraining and protecting. On the one hand, it limits the motion range and the motion mode of the object. On the other hand, with shells, matter and energy of the object can be preserved, and its movement can be sustained.

Shells can be roughly divided into two types, dynamic shells and material shells. The CFVT holds that photons and particles, moving at a speed close to photon velocity, have no shells and interact with everything they come into contact with. Atoms and all composite particles have shells, invisible dynamic shells, that can't be seen by any means. Molecules, cells, and the things they are made of, all have material shells. Shells of molecules and cells are partially visible, such as eggs and metal clumps, but most of them are invisible. The surface layer of celestial bodies, the earth's crust, the outer walls of buildings, the frames of machines, the skins of animals and plants, and so on, are the most conspicuous material shells.

Vortex motions are the most common movements in the universe. They must have original shells to be born. Most of them are closed, but there are a small number of vortexes that are loosely sealed, called leaky bodies, that have obvious jet flows. **Under the restriction of these shells, all their motions are transformed into centripetal movements, and each outer ring of matter is transformed into a shell of all the inner ring matter — the spiral arms are also shells.** As a result, they have a large number of shells, which are the guarantee that all the swirling matter and energy can form round celestial bodies.

The Role of the Shell in the Formation of Stars

Each star originates from its own protonebula. It begins as a random movement of a small amount of nebula dust that takes on an irregular shape. After that, the strongest material flow dominates the movement, forms the core, and turns the movement in one direction, causing the vortex movement to begin. Like a rolling snowball, it rolls the dust of the surrounding large area of the nebula into it, growing and becoming strong. The nebula that has not been rolled into it naturally becomes the boundary of the vortex body at this time, which ensures the continuous movement of the vortex. Vortex motion is a kind of permanent centripetal contraction movement that gradually increases the distance between the vortex body and the boundary, and finally becomes an insurmountable gap with the nebula outside the boundary

—its shell. The star thus forms within this closed shell, becoming the nebula's bright core and moving around the centre of the galaxy.

This is how the Sun was formed. Its shell is the Oort cloud, and its inner edge is the Kuiper belt, and the matter inside the Kuiper Belt sphere is the original nebula of the Sun. In a swirling motion it transformed into glowing plasma, which is the primordial material for everything in the Solar System, including the Earth and all life on it. After the Sun was born, the shell continued to exist and, like the skin of an onion, allowed the sun to make more shells. Other stars, including those a hundred times larger or a hundred times smaller than the Sun, go through the same process.

It is not gravity but the shell that makes all linear motion into curvilinear motion, all centrifugal motion into centripetal motion, and all divergent rotational motion into concentrated convergent vortex motion. It is also an indispensable condition for the evolution of the universe. This can be observed by stirring a cup of tea or coffee.

Although the idea that stars come from nebulae has been widely accepted, the solar system is considered an exception. This source theory of the solar system is described in Hawking's works, *A Brief History of Time* (p.136) and *The Grand Design* (7 The Apparent Miracle). Before him, Asimov described it in his *Guide*, and here is the original.

"Hoyle suggests that the matter blasted into space by a supernova may spread through the galaxies and serve as raw material for the formation of new, 'second-generation' stars, rich in iron and other metallic elements. **Our own sun is probably a second-generation star, much younger than the old stars of some of the dust-free globular clusters.** Those 'first-generation' stars are low in metals and rich in hydrogen. The earth, formed out of the same debris of which the sun was born, is extraordinarily rich in iron—iron which once may have existed at the center of a star that exploded many billions of years ago." (1960, p.67)

NOTE: the main body of all stars is plasma, which is metal — molten alloy.

In this book, other metals, especially heavy metals and radioactive metals, are believed to be the products of stellar phase change. They are mainly produced and preserved by low-temperature small stars and planets, rather than giant stars and supernovae, where all matter is indiscriminate plasma.

The Orbital Velocities of Planets Show the Vortex Effect

The speed of a planet's revolution is its centrifugal force index, an antigravity force. I infer from Newton's third law of mechanics that it equals gravity force, proportional to its mass. That is to say, a massive planet, in order to counteract the proportionally larger gravity, needs a higher revolution speed to avoid the fate of falling toward the Sun. In the solar system, Jupiter's mass is more than two and a half times the mass of all the other planets combined. Even though it is not the nearest planet to the sun, it should have the highest revolution speed, and should also have a significant impact on the motions of other planets, especially the neighbouring Mars and Saturn.

Astronomers provide the average orbital velocity (km/s) of the eight planets around the Sun as follows:

Mercury, 47.4, Venus, 35.0; Earth, 29.8; Mars, 24.1; Jupiter, 13.1; Saturn, 9.7; Uranus, 6.8; and Neptune, 5.4.
(https://nssdc.gsfc.nasa.gov/planetary/factsheet)

From Neptune to Mercury, the velocity increased 8.8 times. If we extend the range to Pluto, its average orbital speed is 4.75km/s, **so the system velocity will increase 10 times! There is no mass factor at all in this calculation, which is a clear violation of Newton's laws. It is also very different from mechanical rotating motion. The latter accelerates step by step, from inside out. It is a typical manifestation of the vortex effect.**

Vortex motion is a funnel-shaped contraction of a planet's own power from outside in, each outer layer wider than the adjacent inner layer. When the outer material and energy contract into the inner layer, the inevitable consequence is the increase of material density and energy concentration. If there is no material contraction movement, the increase of energy is shown in the acceleration of rotational speed. All vortex motions have the phenomenon of a gradual increase in rotation speed from the outside in. If the material and energy resources are abundant, the vortexes will not shrink but will only show the phenomenon of gradually enhanced energy from the outside in with the rotation speed increased accordingly. Some water vortexes show this characteristic, which is why it is difficult for people and ships to escape from the maelstrom.

G. Gamow provides the Titus-Bode rule in his book (p.307), the list of the distance from the sun in terms of earth's distance from the sun (and the ratio of increase in two successive distances):

Mercury,0.387; Venus, 0.723(1.86); Earth, 1.000 (1.38);
Mars, 1.524 (1.52); Planetoids, about 2.7 (1.77)
Jupiter, 5.203 (1.92); Saturn, 9.539 (1,83);
Uranus, 19.191 (2.001); Neptune, 30.07 (1.56)
Pluto, 39.52 (1.31)

Ditto the result of vortex motion, which cannot be formed by gravity. Together with the previous orbital velocity series, they form a perfect vortex structure map of the solar system, a flat gyro rotating swimmingly.

Why can't we use gravity to explain this? The answer is simple: **Gravity can never arise circular motion. Instead, it will terminate all circular & curved motion.** This can be demonstrated by the way magnets attract objects. It always follows the principle of **the shortest path,** drawing objects directly to it rather than making them curve or go in circles. Gravity is proportional to mass. The mass of all the planets is less than 2‰ of the mass of the sun, the sun's gravity is irresistible to the planets, they can not violate the principle of the shortest path of gravity, more unlikely to appear aphelion.

Newton devoted much of his time in *The Principia* II to

criticizing Descartes' theory of vortex. He claimed that the vortex was moving faster inside than outside, and that the pressure was transmitted from step to step infinitely far out. But he could not find such a sustained push, 'active principle' he dubbed it, "Without such a principle it will undoubtedly come to pass that the globe and the inward parts of the vortex, being always propagating their motion to the outward parts, and not receiving any new motion, will gradually move slower and slower, and at last be carried round no longer." (*The Principia*, 380), nor could he explain why the planet orbits were curved instead of straight.

The vortex movement of the Solar System has long ended, but the vortex effect has not been eliminated (We stir a cup of coffee with a spoon to create a vortex. After stopping stirring, the vortex will continue for a time). It has manifested, no longer as shrinking volume, but in the form of gradually enhanced energy from the outside to the inside. This is the most reasonable explanation of the foregoing sequence.

Although there is the phenomenon of centripetal force, the planet will not fall toward the Sun due to the blocking effect of two major forces: the angular momentum obtained from the Sun at the birth of the planet and solar radiation. These are two powerful antigravity, and there is no paradox of action at a distance here.

If you agree with the vortex effect of the solar system above, you will agree with the following earth-moon vortex effect, and the same way to prove it. That is, the following three data: the average orbital radius of the Moon from the ground is 384,400 km, and the average orbital linear velocity it around the Earth is 1.023 km/s; The Geostationary Earth Orbit is about 36,000 km from the ground, and the orbiting speed of the satellite there is about 3.1km/s; The first cosmic velocity of near-earth objects is 7.9km/s, which is nearly eight times higher than the orbital velocity of the moon, an Earth-moon vortex system.

While the Milky Way belongs to a spiral galaxy, astronomical observational data does not fully support CFVT. Astronomers have observed that the speed (about

300 km/s) of stars orbiting the galactic centre at the edge of the galactic disk is indeed much slower than that of stars orbiting nearby at the galactic centre (a star, designated S2 by astronomers, about 15 times the mass of the Sun and 7 times its radius, which was recently found to go around the center with an orbital period of only 15.2 years, has an orbital velocity of more than 5,000 km/s; S4714, another star closer to the galactic center, has an orbital velocity of 24,000 km/s), which accords with vortex theory. But it is much faster than the orbiting speed (about 220 km/s) of the solar system in between, which violates the vortex theory.

The reason for this, I find, is the arm itself spiral motion. This secondary vortex phenomenon, so far, is largely ignored, has also allowed astronomers to theorize about galactic cannibalism. Both of these explanations are discussed in more detail in the following sections.

Shells Are a Stable and Orderly Factor in the Cosmos

According to Newton's law, the planets of the solar system are pulled by centripetal force or gravity of the Sun, so they cannot disperse and fly away. However, like gravity, Newton determined that centripetal force is not an innate force of matter, but an impressed (external) force, and he admitted not knowing its source. According to this book's antigravity theory, the Sun, with its strong radiation movements, repels the planets outward. It's not the Sun's gravity but the shells that prevent planets from dispersing.

Even if the Sun's gravity is extremely strong in the Solar System, planets have eccentricities of large or small, indicating that their centrifugal force is far greater than their centripetal force. If the shell put forth fewer constraints, then they would certainly disperse. Even if the impressed force worked, it would have to work at all times and everywhere to keep the planets from receding, and there could be no such force in the universe. And let the shell do the work, is to let nature take its course, effortless.

For understanding, **consider a stunt cycling show —** motorcycle acrobatics. A few motorcycles on the inner surface of a fixed metal ball move multi-directionally at high-speed. **The centre of the ball is empty, with no central gravitational body, like the Solar System without the Sun,** the motorcycles run as smoothly as on the ground. However, if there was a central attraction, they would flip over the moment.

Now the planets are regarded as the motorcycles, and the planet orbits are regarded as the inner wall of elastic balls. Each planet has its own special sphere. It can also create an elliptical orbit with its own centrifugal force. The whole system appears as a small sphere within a big sphere, and it reaches stability and orderliness by way of its centrifugal force rather than centripetal force or gravity. Other stars and all galaxies have multiple shells. The number of shells is several times higher than the number of celestial bodies and they are packed into space, binding the celestial bodies. At the same time, they acts as a buffer and barrier to collisions between celestial bodies.

Note the fact that **the solar system and its shell, the Oort Cloud, have been moving together around the galactic center, and the two have never been separated. And the Oort cloud and its parent system is undoubtedly such a close relationship, which can be inferred that all celestial bodies are in a stable overall structure of the movement.** The cosmos is like a giant and his cells as independent celestial bodies, with no central gravity to hold them. The cells would continue to grow and change, combining into various tissues and organs (star systems and galaxies) but would never leave the organ membrane or skin (the cosmic shell). This is the secret to how the universe is full of vitality and is simultaneously stable and orderly.

Is the universe bound or unbound? It's a fundamental issue in cosmology. CFVT says yes, it is bound because vortex motion cannot be without boundary conditions; vortex motion is always creating boundaries. This answer also declares that the expansion of the universe is not true, and recognize multicosmology.

Why Can't Planets Run Away?

The centrifugal force is the antigravity force, and the eccentricity ratio is the antigravity index. The bigger the eccentricity ratio of a planet, the more likely it is to free itself of the Sun's gravity and run away.

All planetary orbits have an eccentricity ratio, indicating that the planets are moving with an antigravity greater than gravity. It's the shells that keep them together. Not only are the Oort cloud and the Kuiper belt shells that constrain the planets, but also the orbits of the planets are the shells that constrain the neighbouring inner planets.

Of the eight planets, Mercury has the highest eccentricity at 0.206 — and the strongest dispersion tendency, that Venus prevents (probably a reason for Mercury's precession). Venus is the planet with the lowest eccentricity and the weakest dispersion force.

Venus, Neptune, and Earth are low-eccentricity planets. With their orbits being nearly circular, they can be known as the three stars that maintain the harmony and stability of the solar system. Neptune's 'peacekeeping' is particularly important. Located in the core boundary area of the solar system, **Neptune plays the role of 'garrison commander'**, placing constraints on the separation of several 'separatists' and their flighty behaviour. In between Neptune and Earth is a four-star area of high eccentricity. Pluto, however, is beyond the eight planets and considered an 'anarchist', with its eccentricity being even higher than mercury's.

However, they are interrelated and mutually restricted. Planets with low eccentricity can easily to fall into the sun. It is planets with high eccentricity that have saved them. Mercury's effect on Venus and on Earth is obvious, this can also explain why it is the planet with the greatest centrifugal force.

Here we need to recall the characteristic of vortex motion, which is that the rotation speed of each spiral inner- loop is higher than that of the adjacent outer loop. Each loop naturally becomes the shell of all the inner loops (typical in the solar system), and **the whole is a low-pressure body contracting inward, step by step. Once the system is stabilized, the planet's orbits become the symbols of each loop, thus acting as shells. In fact, this is the inevitable result of vortex movement.**

Why Vortex Force and Shells Frequently Cause Explosions

As discussed earlier, every force causes equal anti-force, they must come in pairs. Vortex force is extremely strong, and the anti-force it causes is not only the strongest but also the most violent one; it's the explosive force.

The shell keeps vortex matter and energy from leaking out. Vortex force makes them contract and the result must be one of the following three kinds, or the three kinds that successively occur. First, phase change, from gas, to liquid, to solid. Second, a black hole. When matter is accelerated to the speed of light, the conversion of mass to energy takes place, and the vortex becomes a black hole. Third, an explosion — our current topics.

Explosions are frequent astronomical phenomena that have become rarer because of the long distance between celestial bodies and the short life span of human beings. **A celestial explosion requires two conditions: continuous contraction force and abundant energy, its mass is not important.** These two conditions are often met by newly born stars, which frequently explode. On the other hand, aging stars do not easily explode because they do not meet these two conditions. The reason for the explosion is not that the thermonuclear reaction of the star stops and collapses under the action of gravity, but that **the strong vortex force makes the vortex shrink again and again, like a powerful and incomparable hand that never stops tightening the clockwork, making it 'unbearable' until it falls apart.**

An explosion is a powerful expansion force and an antigravity force. It temporarily, or permanently, interrupts the contraction movement and absolutely suspends the ultimate goal of the vortex movement to shrink the star into a point.

The explosion has a positive effect. It cools new-born stars, throws off the extra angular momentum, slows their spin rates, and allows giant stars to split into binary stars or clusters, which in turn makes them stable with a longer life span.

Under the action of vortex force, the stars that are separated from the explosion will explode again, producing two kinds of smaller objects successively: luminous dwarfs and nonluminous planets. The sun is a typical example of one that forms the solar system in this way.

Dwarfs explode to create satellites and rings. However, given their small mass, they will flame out after an explosion, displaying an intermittent luminescence. Wood-like planets belong to this category of dwarf star, which I refer to as Jovian stars in this book. Earth-like planets also explode, but such explosions are rare because of the low hydrogen content and weak shrinking force. They cool from a molten state and gradually turn solid, often releasing excess pressure in the form of volcanic eruptions. In the universe, their weight is low, and their status is the highest, because organic life and intelligent creatures can only evolve from them, and the Earth is the best of them.

NOTE: this book holds that wood-like planets in the Solar System are dwarfs. Astronomical observations confirm that they are the same age as the Sun, and they are more likely to be the products of the same solar vortex. The comets, the Kuiper Belt and the nearby Oort Cloud are the stuffs of primitive solar vortex broad tail.

Three Vortex Theories: Descartes', Newton's and CFVT

First, **René Descartes' optical vortex theory.** I think the following two books (synopsis) are well described.

One. "The core doctrine at stake in *The World* is that of mechanism – above anything else, the doctrine that matter is completely inert – and Descartes' aim is to provide a mechanistic cosmology, resting on the basis of quantitative 'laws of nature', and a mechanistic physiology. Among the more fundamental things that he sets out to establish, four are of special significance and novelty. The first is that the stability of planetary orbits and the orbits of their moons can be accounted for on a mechanist basis if we envisage the planets being carried in a sea of fluid matter which takes the form of a vortex. The second is that the propagation of light from the Sun can be explained in terms of the centrifugal effects of its axial rotation. The third is that all vital functions can be accounted for mechanistically. And the fourth is that perceptual cognition can be accounted for, at least to a very large extent, in terms of a mechanistic psycho-physiology ...Descartes' purpose in the last three chapters is to show how the behaviour of light rays can ultimately be explained in terms of his theory of the nature of matter and the three laws of motion. Indeed, the theory of matter turns out to be motivated directly by the requirements of Descartes' physical optics, for the first element makes up those bodies that produce light, namely suns and stars; the second element makes up the medium in which light is propagated, namely the celestial fluid; and those bodies that refract and reflect light, such as the planets, are made up from the third element. Moreover, it is the laws of motion that underpin and explain the laws of refraction and reflection of light, and the accounts of phenomena such as the rainbow and parhelia that are based on these. The laws of motion show us that, given the rotation of the Sun and the matter around it, there is a radial pressure which spreads outwards from the Sun along straight lines from its centre. This pressure is manifested as 'a trembling movement', a property which is 'very suitable for light.' ...Descartes' achievement in the Treatise on Light is twofold. In the first place, his vortex theory explains the stability of planetary orbits in a way that presents an intuitively plausible picture of orbital motion which requires no mysterious forces acting at a distance: the rapid rotation of the Sun at the centre of our solar system, through its resultant centrifugal force, causes the 'pool' of second matter to swirl around it, holding planets in orbits as a whirlpool holds bodies in a circular motion around it. Moreover, it explains this motion in terms of fundamental quantifiable physical notions, namely centrifugal force and the rectilinear tendencies of moving matter. In other words, the heliocentric theory is derived from a very simple theory of matter, three laws of motion, and the notion of a centrifugal force. Secondly, this account also enables Descartes to account for all the known principal properties of light, thereby providing a physical basis for the geometrical optics that he had pursued so fruitfully in the 1620s." (Descartes, 2004, Introduction by S. Gaukroger)

Two. "7. Cartesian Cosmology and Astrophysics
Descartes' vortex theory of planetary motion proved initially to be one of the most influential aspects of Cartesian physics, at least until roughly the mid-eighteenth century. A vortex, for Descartes, is a large circling band of mass particle. In essence, **Descartes' vortex theory attempts to explain celestial phenomena, especially the orbits of the planets or the motions of comets, by situating them (usually at rest) in these large circling bands.** The entire Cartesian plenum, consequently, is comprised of a network or series of separate, interlocking vortices. In our solar system, for example, the matter within

the vortex has formed itself into a set of stratified bands, each lodging a planet, that circle the sun at varying speeds. The minute mass particle that form the vortex bands consist of either the atom-sized, globules (secondary matter) or the 'indefinitely' small debris (primary matter) left over from the impact and fracture of the larger elements; tertiary matter, in contrast, comprises the large, macroscopic material element (Pr III 48-54). This three-part division of matter, along with the three laws of nature, are responsible for all cosmological phenomena in Descartes' system, including gravity. As described in Pr III 140, a planet or comet comes to rest in a vortex band when its radially-directed, outward tendency to flee the center of rotation (i.e., centrifugal force; see Section 6) is balanced by an equal tendency in the minute elements that comprise the vortex ring. If the planet has either a greater or lesser centrifugal tendency than the small elements in a particular vortex, then it will, respectively, either ascend to the next highest vortex (and possibly reach equilibrium with the particles in that band) or be pushed down to the next lowest vortex —and this latter scenario ultimately supplies Descartes' explanation of the phenomenon of gravity, or 'heaviness'. More specifically, Descartes holds that the minute particles that surround the earth account for terrestrial gravity in this same manner (Pr IV 21-27). As for the creation of the vortex system, Descartes reasons that the conserved quantity of motion imparted to the plenum eventually resulted in the present vortex configuration (Pr III 46). God first partitioned the plenum into equal-sized portions, and then placed these bodies into various circular motions that, ultimately, formed the three elements of matter and the vortex systems (see Figure 3)...

On the whole, the vortex theory offered the natural philosopher a highly intuitive model of celestial phenomena that was compatible with the mechanical philosophy. The theory was regarded as superior to Newton's theory of universal gravitation since it did not posit a mysterious, occult quality (gravity) as the cause

of the planetary orbits or the free-fall of terrestrial objects. The vortex theory likewise provided a built-in explanation for the common direction of all planetary orbits. Additionally, the vortex theory allowed Descartes to endorse a form of Copernicanism (i.e., sun-centered world) without running afoul of Church censorship...

In the long run, however, Descartes' vortex theory failed for two fundamental reasons: first, neither Descartes nor his followers ever developed **a systematic mathematical treatment** of the vortex theory that could match the accuracy and predictive scope of the (continuously improving) Newtonian theory; and second, many attempts by Cartesian natural philosophers to test Descartes' various ideas on the dynamics of circularly moving particles (e.g., by using large spinning barrels filled with small particles) did not meet the predictions advanced in the Principles (see Aiton 1972)." (Edward Slowik. *Descartes' Physics.* First published Fri Jul 29, 2005; substantive revision Fri Oct 15, 2021, Stanford Encyclopedia of Philosophy)

Second, **Isaac Newton's fluid vortex theory.** All the information is quoted from his *Principia.* (Note. At that time, the ether was generally understood as a gaseous or liquid substance that filled space, hence Newton's and Descartes' models.)

"Cor. 2. If a globe revolve with a uniform motion about an axis of a given position in a similar and infinite quiescent fluid with an uniform motion, **it will communicate a whirling motion to the fluid like that of a vortex, and that motion will by degrees be propagated onward in infinitum;** and this motion will be increased, continually in every part of the fluid, till the periodical times of the several parts become as the squares of the distances from the centre of the globe.

Cor. 3. Because the inward parts of the vortex are by reason of their greater velocity continually pressing upon and driving forward the external parts, and by that action are perpetually communicating motion to them, and at the same time those exterior parts communicate the same quantity of motion to those that lie still beyond them, and by this action preserve the quantity of their motion continually unchanged, it is plain that the motion is perpetually transferred from the centre to the circumference of the vortex, till it is quite swallowed up and lost in the boundless extent of that circumference. The matter between any two spherical superficies concentrical to the vortex will never be accelerated; because that matter will be always transferring the motion it receives from the matter nearer the centre to that matter which lies nearer the circumference.

Cor. 4. Therefore, in order to continue a vortex in the same state of motion, some **active principle** (Latin: principium activum) is required from which the globe may receive continually the same quantity of motion which it is always communicating to the matter of the vortex. Without such a principle it will undoubtedly come to pass that the globe and the inward parts of the vortex, being always propagating their motion to the outward parts, and not receiving any new motion, will gradually move slower and slower, and at last be carried round no longer." (380)

"Cor. 6. If several globes in given places should constantly revolve with determined velocities about axes given in position, there would arise from them as many vortices going on in infinitum. **For upon the same account that any one globe propagates its motion in infinitum, each globe apart will propagate its own motion in infinitum also; so that every part of the infinite fluid will be agitated with a motion resulting from the actions of all the globes. Therefore the vortices will not be confined by any certain limits, but by degrees run mutually into each other; and by the mutual actions of the vortices on each other, the globes will be perpetually moved from their**

places, as was shewn in the last Corollary; neither can they possibly keep any certain position among themselves, unless some force restrains them. But if those forces, which are constantly impressed upon the globes to continue these motions, should cease, the matter (for the reason ssigned in Cor. 3 and 4) will gradually stop, and cease to move in vortices.

Cor. 7. If a similar fluid be enclosed in **a spherical vessel,** and, by the uniform rotation of a globe in its centre, is driven round in a vortex; and the globe and vessel revolve the same way about the same axis, and their periodical times be as the squares of the semi-diameters; the parts of the fluid will not go on in their motions without acceleration or retardation, till their periodical times are as the squares of their distances from the centre of the vortex. No constitution of a vortex can be permanent but this...

COR. 11. If the vessel and the fluid are quiescent, and the globe revolves with an uniform motion, **that motion will be propagated by degrees through the whole fluid to the vessel, and the vessel will be carried round by it,** unless violently detained; and the fluid and the vessel will be continually accelerated till their periodic times become equal to the periodic times of the globe. If the vessel be either withheld by some force, or revolve with any constant and uniform motion, the medium will come by little and little to the state of motion defined in Cor. 8, 9, 10, nor will it ever persevere in any other state. But if then the forces, by which the globe and vessel revolve with certain motions, should cease, and the whole system be left to act according to the mechanical laws, the vessel and globe, by means of the intervening fluid, will act upon each other, and will continue to propagate their motions through the fluid to each other, **till their periodic times become equal among themselves, and the whole system revolves together like one solid body."** (381)

"In all these reasonings **I suppose the fluid to consist of matter of uniform density and fluidity**; I mean, that the fluid is such, that a globe placed any where therein may propagate with the same motion of its own, at distances from itself continually equal, similar and equal motions in the fluid in the same interval of time. **The matter by its circular motion endeavours to recede from the axis of the vortex, and therefore presses all the matter that lies beyond.** This pressure makes the attrition greater, and the separation of the parts more difficult; and by consequence diminishes the fluidity of the matter...

I have endeavoured in this Proposition to investigate the properties of vortexes, that I might find whether the celestial phenomena can be explained by them; for the phenomenon is this, that the periodic times of the planets revolving about Jupiter are in the sesquiplicate ratio of their distances from Jupiter's centre; and the same rule obtains also among the planets that revolve about the sun. And these rules obtain also with the greatest accuracy, as far as has been yet discovered by astronomical observation. **Therefore if those planets are carried round in vortexes revolving about Jupiter and the sun, the vortexes must revolve according to that law.** But here we found the periodic times of the parts of the vortex to be in the duplicate ratio of the distances from the centre of motion; and this ratio cannot be diminished and reduced to the sesquiplicate, unless either the matter of the vortex be more fluid the farther it is from the centre, or the resistance arising from the want of lubricity in the parts of the fluid should, as the velocity with which the parts of the fluid are separated goes on increasing, be augmented with it in a greater ratio than that in which the velocity increases. But neither of these suppositions seem reasonable. The more gross and less fluid parts will tend to the circumference, unless they are heavy towards the centre." (383)

SUMMARY
1. **Boundary, the vessel.** It is actually the entire solar

system bounded by Saturn's orbit. "Cor. Hence the common centre of gravity of the earth, the sun, and all the planets, is to be esteemed the centre of the world." (409)
2. **The vessel is filled with liquid, ether.** It is the medium for transmitting forces between celestial bodies.
3. **The active principle.** It's the only energy that emanates from the core of the solar system, transmitting from near to far out.
4. Results: The heavy objects are concentrated centripetal, while the light objects are arranged outward according to the law inversely proportional to the core distance. Under the constraints of the vessel, the goal of 'the whole system revolves together like one solid body' is finally achieved.

In my opinion, whoever establishes the theory of **celestial vortex must answer three questions.**
1. The nature of vortex. Is it material, or energy, or a mixture of both?
2. Its motion pattern. Is it open or closed, that is, does it diverge outward, or converge inward, or keep balance all the time?
3. Its impetus. Is it powered by something external, or self- propelled?

The answers offered by CFVT is clear. A celestial vortex consists of two types of particles and plasma. One is energy particles, namely free photons, free electrons and neutrinos. They are characteristically invisible, massless, exist in the form of fields, travel in the form of waves, and move at the speed of light, with energy provided by themselves. They are energy, not matter. The other is mass particles, e.g., atoms, molecules, and their combined blocks. They are characterized by being tangible, massive, and exist in discrete forms. Most of them are observable and quantifiable. Their speed is lower than that of the former, the energy of motion is provided by the former — They have energy of their own but can't be used. Plasma is a mixture of energy and matter. It is characterized by being tangible, massive, and exist in discrete forms. Most of them are observable and quantifiable. Their motion & vibration speed is lower than that of energy particles, but higher than that of mass particles, and the impetus of motion is provided half by energy particles half by itself.

Energy particles come from the radiation of matter, mainly from of the contraction of the vortex. When they meet with matter, their energy part combines with matter and becomes new matter. The other part is transformed into the acceleration of material movement. They are the driving force of all vortexes, and there is no any external force. All vortex movements are inward convergent acceleration movements, which force all materials and energy of vortex bodies to gather at their cores, forming black holes and new celestial bodies.

Concerning Descartes' system. He divided matter into three categories by weight, forming three interlocking vortexes and stratified bands. The planets move close to the sun. The minute elements form a ring around the edge of each vortex. They are positioning and rotating by the centrifugal force originated from solar radial pressure. He didn't say whether the system was open to the outside or converging inwards, defaulting to a stable system, the actual solar system.

We now know that radial pressure is indeed a centrifugal force, moving in a straight line, but relatively weak. It can keep planets and comets from falling to the Sun, but not enough to form vortex and keep planets and comets in their stable orbits, unless, as yet, they are bound by their shells. Despite this and other disadvantages, Descartes basically answered the three conditions of vortex theory and conceived a unique model.

Newton described the vortex of celestial bodies in terms of **Hydrostatics, which must be outward diffusion, not**

inward convergence, because the fluid is not contractible. This is stipulated by the definition and the nature of fluid.

"THE DEFINITION OF A FLUID.
A fluid is any body whose parts yield to any force impressed on it, by yielding, are easily moved among themselves." (*The Principia*, 282)

"PROPOSITION XLII. THEOREM XXXIII.
All motion propagated through a fluid diverges from a rectilinear progress into the unmoved spaces." (Ibid.,359)

That is, even if it is gravity force that impresses to the vortex, it will spread out, not attracting in. Newton's universe must be expanding outward and diverging, contrary to the law of gravity. The vortex has a tremendous mass and should be a powerful gravitational body, but Newton never referred to gravity when discussing the subject of vortex. There seems to be no gravity in liquid.

Planets in the vortex of fluid are also 'endeavours to recede from the axis of the vortex.' But the fluid has resistance, and if the density of the planet is the same as that of the fluid, it will move synchronously with the fluid; If it is heavy, denser, it will get more resistant from fluid, and tend to the center; On the contrary, if its less dense and has less resistance, it will be pushed to the outer edge of the vortex. Actually, this is not the case, and the pattern of planets orbiting the sun is also inconsistent with that of Jupiter satellites orbiting Jupiter. Newton therefore concluded that the motion of the solar system could not be explained by vortex theory.

"SCHOLIUM. Hence it is manifest that the planets are not carried round in corporeal vortices; for, according to the Copernican hypothesis, the planets going round the sun revolve in ellipses, having the sun in their common focus; and by radii drawn to the sun describe areas proportional to the times. But now the parts of a vortex can never revolve with such a motion...**so that the hypothesis of vortices is utterly irreconcilable with astronomical phenomena,** and rather serves to perplex than explain the heavenly motions." (385)

The solar system is a confrontation between centrifugal force and centripetal force.

"And the same thing is to be understood of **all bodies, revolved in any orbits. They all endeavour to recede from the centres of their orbits;** and were it not for the opposition of a contrary force which restrains them to, and detains them in their orbits, which I therefore call centripetal, would fly off in right lines, with an uniform motion." (Ibid.,3)

In his model, the liquid is the medium, responsible for the transmission of forces. It will not and cannot leave the solar system, that is, without centrifugal force, because it exists in a vessel. It also has no centripetal force, since it fills the globe with the same pressure everywhere (Ibid.,338). Newton weighed heavily on phenomena, he didn't appreciate the ether hypothesis as well as any hypothesis. This raises the question: what are the phenomena of the liquid and the vessel? From *The Principia*, I can't find the answer.

In Cor.4, Newton failed to answer a big problem: Where does 'the active principle' come from? Without it, motions can not be transmitted outward. I think, the default answer is The Supreme God, a Being eternal, infinite, absolutely perfect. (Ibid., 528)

In *The Principia* (11), Newton cited the rotating-bucket experiment to support his notion of absolute space as the reference frame for all motion. This experiment could be extended to his fluid vortex theory.

NOTE. In the first edition of this book, I claimed that

Newton constructed his fluid vortex theory to disprove Descartes, which obviously misinterpreted Newton. "I shall now add some things relating to the forces of progressive bodies, and to the density and resistance of those mediums in which the motions hitherto treated of, and those akin to them, are performed." (Ibid.,282) That is his real intention. He didn't mention Descartes' name at all in *The Principia*, nor did he completely deny the vortex hypothesis, but said "The hypothesis of vortices is pressed with many difficulties." (Ibid.,527)

I apologize to my readers for this careless mistake. But that doesn't change my critical attitude toward his vortex theory. His approach to setting the boundary of the vortex, the 'vessel wall', should be affirmed, which is what CFVT calls a shell. All vortexes must have a shell, and often multiple shells. But the other features of the vortex he described are unacceptable to CFVT. I regard it as a new paradox created by Newton.

Orion: The Baby Star Garden

At the meeting of the American Astronomical Society on June 9th 2009, Professor Charles Townes of the University of California, disclosed in a statement: "New measurements have found that **the diameter of Betelgeuse has decreased by 15% in the past 15 years, with a gradual but accelerating trend year by year.**"

The consensus among astronomers is that Betelgeuse is only eight million years old. It is massive and spinning at a remarkable rate of fifty-four thousand kilometres per hour. That's fifteen thousand meters per second! It's a red supergiant ready to explode as a supernova and collapse into a neutron star, but what it's doing right now is obviously not collapsing. Although Betelgeuse is shrinking, there is no significant change in its brightness. At the current rate of contraction, its electron degeneracy will not be destroyed, and its core will not become a neutron star. Professor Townes and his team are baffled by this phenomenon.

According to a HR diagram, giant stars like Betelgeuse are short-lived due to their rapid energy and material consumption. The larger the mass, the shorter the life — this is a firm belief among astronomers. Although young, these stars in Orion will soon die in violent and short collapses, each leaving a neutron star as a tombstone.

In my opinion, there are two obvious phenomena: **high spin-speed and sharp contraction are impossible to appear on aged or dying celestial bodies**. From the available data, it can be determined that Alnitak, Alnilam, and Mintaka are very young new-born stars with low density, so they emit red light. They continue their swirling motion, cause self- contraction and explosion, which had made Alnitak split into three stars not long ago. Betelgeuse is the slowest of the cluster, and it has just entered the contraction phase.

Compared to the 5-billion-year-old sun, Betelgeuse is a typical baby star. From the perspective of CFVT, its current contraction is inevitable, the rate of contraction will accelerate, and quite likely to shrink by several times from the current volume.

The terminal of celestial contraction is a black hole, which makes Betelgeuse look unstable, and what we observe is that its brightness fluctuates. Astronomers have observed this phenomenon.

On February 28, 2020, Nature magazine published Davide Castelvecchi's reports entitled "Mysterious Faded Star Betelgeuse Has Started to Brighten Again — 'Orion's shoulder' had reached unprecedented dimness in mid-February, leaving astronomers befuddled." The abstract of the article is as follows.

"Easily recognizable as the right 'shoulder' in the constellation Orion, Betelgeuse is normally one of the ten brightest stars in the night sky. But it began getting dimmer in October last year, and by mid-February it had lost more than two-thirds of its brilliance — a difference noticeable to the naked eye. But the star has now brightened by around 10% from its dimmest point, says Edward Guinan, an astrophysicist at Villanova University outside Philadelphia, Pennsylvania, whose team has been tracking it for 25 years." (https://www.nature.com/articles/)

From this report, I judge that the central black hole of Betelgeuse has been formed. Its brightness fluctuations are caused by it. Over the past few decades, the black hole has had plenty of food to keep it bright. As the black hole grows, it eats material from within at an accelerated rate, leading to an internal ring of isolation in October 2019, where the black hole's food supply plummets and its brightness decreases correspondingly. However, in the overall powerful contraction, this inner ring is soon eliminated, the black hole is fed again, and the stars begin to brightened up. This brightness fluctuation will appear repeatedly, and the cycle will be shorter and shorter. Finally, 99% of the stellar stuffs will fall into the black hole, forming a big one, Betelgeuse may become a shadow or disappear completely for a short time in the future. Here is its precedent.

"IT'S A STELLAR MYSTERY: HOW DID THIS GIANT STAR SUDDENLY DISAPPEAR FROM THE SKY?
Astronomers are torn between one of two scenarios. ESO PASSANT RABIE 6.30.2020 6:00 PM
For ten years, astronomers had been observing a giant star located in a dwarf galaxy that's 75 million light-years away. The luminous blue variable star was one of the largest of the known universe, and about 2.5 million times brighter than the Sun. However, one day, the star suddenly vanished, leaving no visible trace behind."
(https://www.inverse.com/science/disappearing-star)

'Giant star pulls off vanishing act. Did it become a black hole or was it all an illusion?' By Elizabeth Howell published June 30, 2020. (https://www.space.com/)

As the observation progresses, astronomers will find that this is one of the phenomena that are not uncommon in the starry sky. A black hole is a transitional celestial body, or the fetal form of a star.

After going through a relatively brief black hole stage, Betelgeuse will evolve from the current ionic state to a higher density plasma, then explode into several blue stars, which makes it absolutely impossible to produce a neutron star in its centre — there is no star like this in the universe.

There are many such celestial bodies in the Orion Nebula, that we could call it a thriving garden for baby stars.

A Sky of Fire and Ice

Ice is the solid state of water. In our daily life, we often use water to extinguish fires, so there is an idiom of 'incompatibility between water and fire.' But in physics, they have one thing in common: their motion velocity & vibration frequency, which is positively correlated with temperature. We know, temperature is nothing but index of corpuscular motions & vibration. They move & vibrate fast, and when we observe, it is high temperature; move & vibrate slowly, low temperature. No any particle motion & vibrate, no temperature, called absolute zero. The motion and vibration of corpuscles are causes and the temperature is the effect.

Here, I refer to ions, molecules, atoms, protons, neutrons, electrons, photons, neutrinos, etc., as corpuscles in general, and they constitute various substances, among which the largest number is plasma. Corpuscles are moving & vibrating at all times, but the velocity and intensity are different, depending on their interaction with things, which shows temperature fluctuation.

"The volume was now seen to depend on the velocity of the molecules. The higher the temperature, the faster they move, the more 'elbow room' they require, and the greater the volume. Conversely, the lower the temperature, the more slowly they move, the less room they require, and the smaller the volume." (Asimov, 1993)

The motion of any object requires space and is positively related to the velocity of the object, and corpuscle motion is no exception. **As the corpuscles move & vibrate faster, they need more space, so the object must expand, its density drops, its temperature rises, its solid phase changes into liquid and gas, some particles move closely** to light velocity, and they changes into ions and photons, which are fire and light. On the contrary, if the corpuscles slow down, it must shrink the object, increase its density, decrease its temperature, and undergo the opposite phase transition, turning the gas into a liquid and a solid, and water into ice. Fire and ice are the two extremes of corpuscular motion, and these two incompatible things are thus united.

In this way, the opposing concepts of cold and hot are unified. Cold is the relatively slow motion & vibration of corpuscles, and heat is their relatively fast motion & vibration. Dim and cold bodies such as planets, moons, stardust, and so on, can also be united with glowing, hot bodies, chiefly stars. They all are plasma, composed of relatively low-speed ions and relatively high-speed ions, respectively. The source of all things in the universe is thus unified.

Ninety-nine percent of the universe is plasma, however most of the objects we observe are not ionic structures, but atomic and molecular structures that are created when ions slow down. This is why atmospheres, water, minerals, and organisms appear on planets rather than stars. And yet planets are bursts of stars, and the source of everything is plasma.

Along this line of thought, the development of the universe and the process of creation can be clearly described. That's the holistic view of the universe — the philosopher's methodology. From this new horizon, we have gained a new understanding of many popular ideas. Here are three of them:

First, blazing core theory of stars. The Big Bang occurs from a singularity, and the core temperature is the highest. Nuclear fusion takes place in the core area of every star, which also forms high temperature. **From the point of view that corpuscular velocity determines temperature, the situation is just the opposite. The temperature of stars decreases gradually from the outside in, and the core temperature is the lowest. It can also explain the high**

temperature in the ionosphere, corona and 'fire wall' on the edge of solar system, though they are not hot due to their low density.

Second, the theory of the high-temperature formation of elements, especially heavy elements. Scientists generally believe that metals require high pressure and high temperatures to produce them. The metals on planets like Earth come from the Sun, and the metals on the Sun are formed by supernova explosions.

According to CFT, on the other hand, **most metals are formed when the velocity & vibration of particles are flat, viz., at relatively low temperatures,** whereas the high temperatures only melt them into a single mass, a plasma.

And finally, the star collapse and death theory. The popular theory is that when old stars burn out of nuclear fuel (hydrogen) and fusion stops, the star collapses into a white dwarf, a neutron star, or a black hole, and goes into a state of decay. According to CFT, **the core of a star is the densest and the coldest part of the star without fusion, so it doesn't collapse, and ions won't age, the stars are eternal.** Stars die when the spiral arms roll them into the galactic black hole, but this is the metabolic form of the universe, not the actual death of stars.

The most striking feature of holism, as compared with current corpuscular theory, is that it focuses on the whole picture and ignores the details. **The most important factors for vortex motion are the mass, energy, mode, and motion of objects.** As to vortex force, the four forces retire; the size, strength, rotation direction and color of the particles are meaningless and can be ignored. And that phlogiston, bosons, virtual photon and other messenger particles all are dispelled.

Firelight: Fire Is Different from Light

Is not light grander than fire?
                                   Thomas Carlyle

The burning phenomenon often seen in daily life is accompanied by firelight. However, fire is not light, although fire usually gives off light — there are many unlit fires, and many lights that have no fire.

The differences between fire and light are as follows. First, motion velocity. **The fire fluttered in place, but the light drifted outward at the speed of light.** Fire does not fly, it only spreads, but it does emit light all over the body and transmits its own information, and information of combustion products, by means of light. In other words, fire has no independence. It cannot be separated from burning materials. Things burn out and the fire extinguishing, but light is completely free.

Second, their properties. **Fire is plasma matter; light is energy and field state.** Fire itself has a high temperature, but light does not have any temperature, excepting it contact and interaction with things.

Third, their conditions. Fire is the aggregation of matter, and decomposition is the premise of aggregation.

If there is only decomposition and no aggregation, no fire. Like splitting water molecules into hydrogen and oxygen atoms; break an atom down into ions, they don't get fire. The aggregation of matter itself will not produce fire too. For example, two hydrogen atoms fuse to form a molecule of hydrogen, and two hydrogen atoms and one oxygen atom fuse to form a molecule of water, gold and silver atoms fuse into gold nugget and silver nugget, the rain converges into rivers and lakes, etc. Stars are made of pure plasma, they are only luminous, they do not burn or ignite. From here, we can see that the conditions for luminescence are much broader than the conditions for fire. Both decomposition and aggregation of matter can emit light; electric light, phosphorescent light, and auroras are of this class, with starlight being the strongest and most common one among them.

Fourth, the results. **Fire is a chemical phenomenon.** After burning, objects are transformed into other substances. Prototypes change or disappear and are always visible. Light is a product of mass-energy conversion, a physical phenomenon. **Light will enter the light field immediately after escaping from fire.** It is visible only when interacting with objects.

These differences between fire and light have not been recognized by many people so far. So it is not surprising that even Isaac Newton regarded the two as one. In *Principia*, he mentioned that "To this purpose the philosophers say that Nature does nothing in vain, and more is in vain when less will serve; for Nature is pleased with simplicity, and affects not the pomp of superfluous causes." (387) He listed several facts as proof, one of which was that the causes of 'the light of our culinary fire and of the sun' (387) are the same. Now, we know that the causes of the two are different. The light of **the culinary fire comes from burning of combustible substances, its main body is fire; while sunlight is caused by radiation and resonance of plasma. Its main body is light, not fire**—Under the action of the light barrier, the corona region will produce material including oxygen and cause combustion.

Combustion is a process of accelerating the movement of fuel particles, which is generally divided into three stages: oxidation, ionization and light. Take coal combustion as an example.

One, we ignite the coal and accelerate its movement of coal molecules and air molecules. When the temperature rises to 950 degrees Celsius, the oxygen ions (Of all the elements, it is the second most electronegative) in the air will take away the electrons in the coal, and,

Two, the coal become a plasma, that is, coke. The plasma glows and does not burn, but it triggers combustion of the air in its immediate vicinity, a chemical reaction, forming a hot shell that prevents the plasma from dispersing.

Third, the plasma temperature continues to rise, in a slow and continuous way to break through the hot shell, as free ions and photons fly apart.

Common fuels are complex in composition and the three stages of their combustion often occur intermingling. In addition, partial combustion would not be hot enough to ionize the fuel, which would be lost in the air as smoke.

Fire is discrete and its parts have different temperatures, colors, and shapes. The light is one, which makes it holographic. This feature allows us to see the fire scene in its entirety from the tiny amount of light that enters our eyes. Newspaper reporters often describe a fire as a 'firestorm', but this is a literary term, not a scientific description. From the latter point of view, what he saw was not the real thing, not the fire, but the image transmitted to him by light. He was actually experiencing a special event of light wave propagation that was no different from watching a movie in a theater. This includes a wide spectrum of things, from the burning of a candle to the 'burning' of the Sun and all the stars, and all that we can see with our eyes is their image — by the light that is emitted through them. Moreover, we are often kept at a safe distance, long or short, from the scene of the fire, so all we see are historical images.

In *Three Roads to Quantum Gravity*, American theoretical physicist Lee Smolin has an interesting description that helps us understand the phenomenon.

"We are very used to imagining that we see a three-dimensional world when we look around ourselves. But is this really true? If we keep in mind that what we see is the result of photons impinging on our eyes, it is possible to imagine our view of the world in quite a different way. **Look**

around and imagine that you see each object as a consequence of photons having just travelled from it to you. Each object you see is the result of a process by which information travelled to you in the shape of a collection of photons. The farther away the object is, the longer it took the photons to travel to you. So, when you look around you do not see space—instead, you are looking back through the history of the universe. What you are seeing is a slice through the history of the world. **Everything you see is a bit of information brought to you by a process which is a small part of that history."** (2007)

The Sun Shines but There's No Fire—

No Nuclear Fusion

Why does the Sun shine? It is generally believed to be the result of nuclear fusion. Under the high temperature and pressure of the Sun's core, four hydrogen nuclei are in effect converted into one helium nucleus, a fraction of the mass being released as energy, called Bethe-Weizsäcker-cycle. Later, Carl Friedrich von Weizsäcker, the German physicist and philosopher, and Hans Albrecht Bethe, the German-born American physicist, improved it to CNO cycle, viz., carbon-nitrogen-oxygen cycle. **They were trying to decipher the enigma of star power, and also they wanted to elucidate the direct correlation of energy production and the development of the heavy elements in stars.**

Anyhow, it is important to understand that high temperature and high pressure have double effects. It promotes both polymerization and decomposition, and the latter is more common, especially in high temperatures. Even if hydrogen can be fused into helium in the Sun's core, it will immediately reduce back to hydrogen ion.

In fact, under such high temperatures and pressures, there are only plasma in sun, a collection of the two simplest ions of the element table, and no atoms and molecules. Can we find them in the molten steel of a blast furnace? In the Sun, all atoms, if they can be formed, must be short-lived and cannot be preserved or accumulated. And since radiation takes a small share of the total energy away, the sun loses only a small amount of energy during its life, so that its temperature and pressure do not drop significantly, and the state of plasma can always be retained. The idea that the hydrogen in the Sun's core would coalesce into helium, helium would coalesce into lithium, and eventually into carbon and iron, decaying into white dwarfs, clarifies that the Sun's decline is untrue and impossible.

The Sun is a giant ball of plasma, which is composed of semi-trapped photons, and it must emit light and heat. This is how the molten steel and lava glow. They are all strongly radiating objects that don't need thermonuclear reaction or nuclear fusion—Would you believe that lava is nuclear fusion?

Yet, that the astronomers claimed that the two major components of the Sun are hydrogen, ¾ and Helium, ¼ seem to be evidence of nuclear fusion inside the Sun.

The answer can be found in *Asimov Guide*.

"As early as 1816, an English physician named William Prout had suggested that all atoms were built up from the hydrogen atom."

"The most common materials in the universe generally are hydrogen and helium. **Hydrogen atoms make up about 90 percent of all the atoms there are, and helium atoms make up another 9 percent. This fact may not be surprising when one considers that hydrogen atoms are the simplest in existence, with helium atoms second simplest.** Of the atoms that remain, carbon, oxygen, nitrogen, neon, and sulfur make up the bulk. Hydrogen and oxygen atoms combine to form water molecules; hydrogen and carbon atoms combine to form methane molecules; hydrogen and nitrogen atoms combine to form ammonia molecules."

The sun doesn't allow elements more complex than helium, let alone water, methane, ammonia, etc. We can also ignore their element tags and just call them **simplest & second simplest ions,** which is easier to understand. At high temperatures, there are no atoms, no molecules, few complex ions, just the two simplest ions.

That is to say, even for ions, the complexity of its structure is inversely proportional to the temperature, and **the hotter a star is, the simpler the ions are.** That is why stars are mostly hydrogen and helium, and blue giants have the highest hydrogen ratio.

The Formation of the Sun and the Solar System

"The nebular hypothesis says that the Solar System formed from the gravitational collapse of a fragment of a giant molecular cloud. The cloud was about 20 parsec (65 light years) across, while the fragments were roughly 1 parsec (three and a quarter light-years) across. The further collapse of the fragments led to the formation of dense cores 0.01–0.1 parsec (2,000–20,000 AU) in size. One of these collapsing fragments (known as the presolar nebula) formed what became the Solar System. The composition of this region with a mass just over that of the Sun (M☉) was about the same as that of the Sun today, with hydrogen, along with helium and trace amounts of lithium produced by Big Bang nucleosynthesis, forming about 98% of its mass. The remaining 2% of the mass consisted of heavier elements that were created by nucleosynthesis in earlier generations of stars. Late in the life of these stars, they ejected heavier elements into the interstellar medium."
(https://en.wikipedia.org/wiki/Formation_and_evolution_of_theSolarSystem)

This 'giant molecular cloud' is nothing but the part of Oort Cloud within the Kuiper Belt. Its diameter is not as wide as above hypothesis, only about 60 AU, 9 light hours. Its original diameter is about 250 AU, 33 light hours. And it's not a molecule cloud, but a particle cloud; Not gravitational collapsed, but the product of vortex motion. We can imagine that six billion years or more ago, like the rest of the Oort Cloud, the space of the present solar system was full at that time, and that solar vortex motion took place in the centre of it, where the sun is now. Constantly swirling the surrounding matter, the vortex gradually expanded, reaching a radius of 122 AU at its peak, with the sphere radius of the

Kuiper Belt having been identified by Voyager 2 in 2018.

The nebula's matter, which was lightless, began to glow when the main body of the vortex shrank into a red giant — a dim gaseous plasma. **I assume that Neptune's orbit was the boundary of the solar red giant, and that it had a radius of 30 AU, similar to the original Betelgeuse, and that it continued to spin and contract at high speed.**

One of the characteristics of vortex motion is that its core particles move at the highest speed. When they reach the speed of light, the mass-energy conversion occurred, and the core black hole emerged. As the material of the solar red giant continuously rolled into the black hole, at a critical point, energy-mass conversion occurred, and the black hole instantly became a visible star, thus our Sun was born. It accounted for more than 99% of all vortex materials, and the nebula tail, which had not been coiled into the black hole, formed Jovian stars, comets and KBOs (Kuiper belt objects), all of which revolved around the Sun.

The new-born Sun was a high-speed spinning blue star, several times larger than its black hole, shrinking close to its current size about one million years later. The over-pressurized energy caused the Sun to explode, creating six small suns: five Earth-like planets and the Moon. The fifth planet, beyond Mars' orbit, smashed into a Jovian star. The reason for the collision was that there was no room for Earth-like planets around the Sun, and now Mercury orbits the fifth Jovian star. The explosion of the sun pushed the whole Jovian star out, creating space for Earth-like planets. At the intersection, the fifth planet and the fifth Jovian star collided, and their fragments formed the asteroid belt.

Jovian stars are directly composed of Oort cloud material, are mixtures of gaseous and liquid plasma, characterized by high mass, low density and high spin-speed. The mass-energy conversion and energy-mass conversion motions occur alternately, and they become small stars with intermittent luminescence. Earth-like planets are composed of liquid plasma of the Sun, with

relatively small mass and high density, and energy-mass conversion is only carried out in one direction. They contracted to coalesce into gaseous, liquid, and solid atomic materials to form atmospheres, oceans, land, and mineral deposits, which are now the four Earth-like planets and the moon. Earth is the most superior of these planets as animals, plants and humans have evolved on it.

Through this explosion, the Sun lost 1% in weight and lost more than 98% in angular momentum, entering the long and stable period of the main sequence star, and eliminating the risk of re-explosion.

This is a more realistic picture of the solar system, which is not exactly what *An Outline of Force Cosmology* described. With the deepening of astronomical observation, the current description cannot guarantee accuracy. What I can confirm is that the fundamental principles remain the same, the cosmic force, vortex motion and their motion patterns never change, which is completely contrary to standard cosmology.

The Sun is thought to be mainly composed of hydrogen and helium, but I simply think it is a plasma, which emits light by itself and does not depend on thermonuclear reaction or nuclear fusion to glow. **Accordingly, it doesn't have a series of atomic transmutation movements, and it won't turn into a red giant star, white dwarfs, and black dwarfs one after another in the later period. It was always a glowing plasma before entering the black hole of the galaxy, and this is also the case with other stars, which is my very certain belief.**

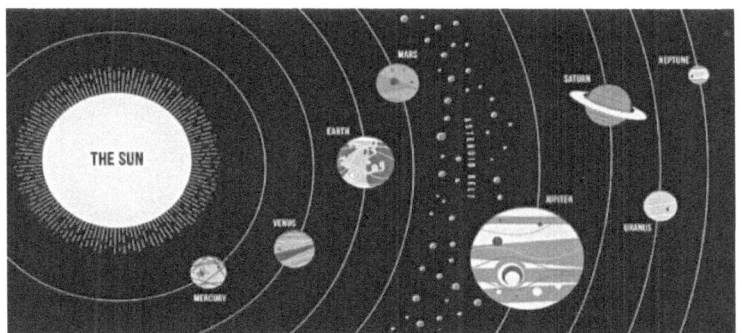

Online public picture

TIP. Comparison of the Sun's volume in red giant stage with that in main sequence:

The formula for calculating the volume of a sphere: $V_0 = (4/3)\pi R^3 = 4.1889 R^3$
It is known that the radius of the Sun, $R = 696295$ km; $V_1 = 1.4 \times 10^{18}$ km^3.
Red giant sun radius is 30 AU, that is, the space from the present Sun to Neptune orbit,
$R = 30 \times 1.5 \times 10^8$ km $= 4.5 \times 10^9$ km.
$V_2 = (4/3)\pi \times (45 \times 10^8 \text{ km})^3 = 381540 \times 10^{24}$ km^3
$= 3.8 \times 10^{29}$ km^3.
The result of comparison,
$V_2 : V_1 = 3.8 \times 10^{29}$ km^3 : 1.4×10^{18} km^3 $= 2.7 \times 10^{11} : 1$

The red giant sun is 2.7×10^{11} times larger than the present sun, or 270 billion times larger; it underwent about 38 doubled contractions to reach its current volume.

S_1, 270000000000/2 ➤ S_2, 135000000000/2 ➤ S_3, 67500000000/2
➤ S_4, 33750000000/2 ➤ S_5, 16875000000/2 ➤ S_6, 8437500000/2
➤ S_7, 4218750000/2 ➤ S_8, 2109375000/2 ➤ S_9, 1054687500/2
➤ S_{10}, 527343750/2 ➤ S_{11}, 263671875/2 ➤ S_{12}, 131835938/2
➤ S_{13}, 65918000/2 ➤ S_{14}, 32959000/2 ➤ S_{15}, 16479500/2

- S_{16}, 8239750/2 ➤ S_{17}, 4119875/2 ➤ S_{18}, 2060000/2
- S_{19}, 1030000/2 ➤ S_{20}, 515000/2 ➤ S_{21}, 257500/2
- S_{22}, 128750/2 ➤ S_{23}, 64375/2 ➤ S_{24}, 32188/2
- S_{25}, 16093/2 ➤ S_{26}, 8047/2 ➤ S_{27}, 4023/2
- S_{28}, 2011/2 ➤ S_{29}, 1006/2 ➤ S_{30}, 503/2
- S_{31}, 251/2 ➤ S_{32}, 126/2 ➤ S_{33}, 63/2
- S_{34}, 32/2 ➤ S_{35}, 16/2 ➤ S_{36}, 8/2
- S_{37}, 4/2 ➤ S_{38}, 2/2 =1

Now import the black hole, and this sequence will be shortened to 11.

S_1-red giant sun, 270000000000/2 ➤ S_2, 135000000000/2
➤ S_3, 67500000000/2 ➤ S_4, 33750000000/2 ➤ S_5, 16875000000/2
➤ S_6, 8437500000 ➤ S_7, 4218750000/2 ➤ S_8, 2109375000
➤ S_9-black hole sun ➤ S_{10}-blue giant sun, 4/2
➤ S_{11}-main sequence sun, 2/2 =1

And we can use this method to observe the evolution of Betelgeuse and other red giants.

This account is so speculative that some readers may see it implausible, and my defense is as follows.

Galaxies are made up of clusters of stars, which outnumber galaxies by several orders of magnitude, and star black holes outnumber galactic black holes in the same proportion. Due to their relatively small size and enveloping by the crusts of red giant all the time, no one has ever seen any of them.

It can be concluded that **we can never directly observe a stellar black hole.** This is especially true for smaller black holes and most other astronomical phenomena. Most of what we can observe are events hundreds of millions of years ago, their entities may no longer exist. **Most of popular cosmological theories are thus based on guessing and conjecturing.**

What makes speculative philosophy superior to guess and conjecture lies in its logical process. It starts with the establishment of the First Principle, obeys the law of cause and effect, and unfolds problems step by step from top to bottom, just like the Ariadne thread, which guides explorers to a certain exit. **Speculative philosophy sees the First Principle or First Cause as the Torch in a cave, without it, all our explorations would have to grope in the dark.**

The Mechanism that Forces a Comet to Turn Around

According to Newton's laws, if a comet is attracted by the Sun's gravity, the distance between it and the sun will tend to be zero, and the gravity will tend to be infinity, and the comet must fall toward the Sun. Even it has centrifugal acceleration to avoid falling toward the Sun, there is no reason for the comet to take a sharp turn around the Sun, because gravity has linear rather than curvilinear motion. Please stop here for a minute. **Imagine a magnetic ball the size of Sydney Opera House (the sun). We roll a small steel ball (a comet) beside it. Can it not be firmly attached on the magnetic ball, but turn around and leave quickly? This monstrous phenomenon can only be explained by CFVT.**

Modern astronomers tell us that the Oort cloud is the birthplace of comets and that most of them wander through that vast, cold space with a small number of intruders into the inner solar system. Suppose a comet runs in and its destination is the Sun. When it gets close to the Sun, the sunbeams begin to act on it first, wanting to push it away. As the comet approaches the Sun's surface, the Sun's gravity begins to act on it, wanting to pull it nearer. But at the same time, the radiation gets stronger, and the two forces impinge. The results depend on the mass of the comet. If a comet's mass is big enough, and if its & the Sun's combined gravity ($M1 \times M2$) is greater than the radiant force of the Sun, the comet will drop into the Sun. If it is smaller than that, it will fly by. If the two forces act against each other, the comet will go around and fly back, or become a new object around the Sun. Such objects have existed in the past and will exist in the future, but we fail to see them simply because they drop in sun or burn by sun so easily, or are pushed away from it.

The strong evidence for this is the comet's tail. It is formed by the moisture and dust particles of the comet, blown by the strong solar wind. As the distance between the comet and the Sun shrinks, the tail becomes larger and longer, always extending outward. **This shows that the Sun's radiant force is far greater than its gravitational force. The radiant force is the antigravitational force that blocks comets, and even planets, from falling into sun.**

Some comets — such as Halley's comet, the paragon comet — can be as far away from the Sun as more than one hundred AU, but they will return regularly, which cannot be explained by Newton's law because the Sun's gravity is already negligible. From the moment the comet turns around, it shows that it has successfully gotten free of the Sun's gravity; the farther back it goes, the weaker the Sun's gravitational pull. The weakening sun radiation has always played a repulsive role, and there is no reason for the comet to turn back again, which can only be explained by the role of shells. The two values of the comet's apogee and perihelion differ by more than 1,000 times, and gravity alone cannot explain it.

Not only comets, but also other objects orbiting the sun are forced by shells, which are the main factor that forces the celestial bodies to move in their respective orbits. **Enclosing by shells, none of the objects, even dusts, within the Solar System can escape, and they must move in curves.** When Halley's comet moved toward the Sun, it relied on the concave orbits of Mercury and Venus, and the concave inner side of the corona, to slide past as fast as we do on a roller coaster.

The Gravity of the Sun Is Weaker Than That of the Planets

Following Newton's law, NASA calculated that **the Sun's G value (eq.)(m/s^2) at 274 & the Earth's at 9.78.** (https://nssdc.gsfc.nasa.gov/planetary/factsheet/sunfact.html) That is to say, the surface gravity of the Sun is twenty-eight times the gravity on Earth's surface. So, an object weighing 100 kilograms on Earth will weigh 2800 kilograms on the sun. **This is the result of considering only mass without temperature and density** (specific gravity). The influence of temperature and density on gravity is certain.

For example, when the sun becomes a red giant in the late period and its diameter is one hundred times larger than the current diameter, according to Newton's law its gravity will be inversely proportional to the square of its distance, and its surface gravity will drop 100^2 times. On Earth, an object weighing 100 kilograms will weigh only 280 grams on the sun surface at that time.

A common fact is that heat expands and cold contracts by the changing of the distance between the particles. When an object of definite mass heats up, its volume expands, the distance between particles increases, the object's density decreases, and its surface gravity decreases. On the contrary, when it gets cold and shrinks, the distance between particles decreases, the density increases, and the gravity increases. Not only that, but also the rise in temperature means that the radiation force (tension) increases. It is an antigravity force. The foregoing calculation does not consider radiation force.

Speaking of density, the current density of the sun is about one-quarter of that of Earth, so the 2800-kilogram object should be adjusted to 700 kilograms. Add in the temperature, with the core of the Earth being about 6,000K and the Sun's being 16,000,000K, and we find that the latter number is 2600 times higher than the former number. Coupled with the strong radiation force of sunlight, the actual gravity of the Sun, we believe, is much weaker than that of Earth, meaning it has almost no influence on the motion of the planets and is far inferior to its radiation force in this respect. In my eye, **ionspouts rage on the surface of the sun, and solar wind can blow out long comet tails & push light sails outward, which are all powerful manifestations of antigravity. So, where can gravity stand?**

Here, many shortcomings of the law of gravity are exposed. The law only considers the mass and external distance of objects and does not consider the distance between molecules and atoms within objects, that is, the temperature and density of objects, and is therefore imperfect —Even if only considering mass and external distance, when the sun becomes a red giant, it will overturn the law of gravity. It also lacks the factor of the movement of objects, that is, the difference between objects in phase motion and in opposite motion, and ignores the antigravity factors of radiation and tension, plus irreversible action at a distance. As a basic physical law, its misleading effects must be far greater than its guiding effects.

Note: The above discussion is based on the theory of solar core heat and nuclear fusion theory. And now that I've replaced it with the core cryogenic theory, it's really possible that the Sun's gravity is 28 times stronger than that of the Earth's. **But the fact of the comet motion above suggests that it has no action at a distance, and that its gravity is far less powerful than its radiation force in space.** G is related to escape velocity, which, according to NASA, is considered to be 617.6 km/s in the sun surface, which is obviously not true. The sun is a strong radiator. The speed of the solar wind is 200-800 km/s. Comets are blown out with long tails by it. If things on the surface of the sun are not burned, they will be vaporized, weightless and blown away at high speed, needn't any escape velocity.

How Strong Is the Earth Gravity?

I define gravity as the aggregation force between molecules and atoms. The indispensable condition is mutual contact and interconnection, which absolutely excludes the action at distance. Objects that are not connected to each other or that repel each other can be stable in the same space, that is the function of the shell and external pressure, which is a much stronger force than gravity.

How strong is the terrestrial gravity? The answer I give is 1.293 kg/m^3. That is, 1,293 grams of material evenly distributed in a cubic meter of space above the Earth surface that can be attracted by each square meter of crust, which contains the weight of air above 1 meter. In *The Principia*, Newton called this ratio weight, not gravity.

How did I get this number? Middle school students will recognize it at a glance: 1.293 kg/m^3 is the density of air, also called specific gravity. As a comparison, the density of water is 999.972 kg/m^3, which is simply 1,000 kg/m^3, that is 1 ton per 1 cubic meter.

Gravity is an interaction. If Earth is M1, then the air on a second planet is M2. The air may have a higher density, such as happens when it is mixed with a great deal of liquefied gas, which multiplies the density, so that it is closer to the Earth. If the space is filled with hydrogen, the density of the air drops, and the air rises off the surface, meaning that gravity fails. The figure 1.293 kg/m^2 can be used as the gravity index of Earth, so when we multiply it by the total area of Earth — 510 million square kilometres —we get the total gravity of Earth, which is about 660 billion tons. In *The Principia*, the gravity can also be translated into air density, pressure, or weight.

"Suppose the compressing force to be in a duplicate ratio of the density, and the gravity reciprocally in a duplicate ratio of the distance, and the density will be reciprocally as the distance. To run over all the cases that might be offered would be tedious. But as to our own air, this is certain from experiment, that its density is either accurately, or very nearly at least, as the compressing force; and **therefore the density of the air in the atmosphere of the earth is as the weight of the whole incumbent air,** that is, as the height of the mercury in the barometer." (*The Principia*, 292)

This index does not apply to other planets, because their gravity varies depending on mass, temperature, density, pressure, and spin-speed.

Here I mention only Earth surface gravity and not Earth's centre gravity, which is related to my unique view of gravity that I will explain in the following chapter.

Gravity and Tension

Where is the gravitational centre of the Earth? The most common answer is in the centre point of the Earth. Newton's first definition of centripetal force is "Of this sort is gravity, by which bodies tend to the centre of the earth."(*The Principia*, 3)

According to the tug-of-war mode, this answer is wrong. Because the centre point of the Earth is the midpoint between the two halves of the Earth, just like the equilibrium point between two evenly matched tug-of-war teams, the forces of both sides are canceled out, and the gravity is zero.

According to Newton's law, this answer is also wrong because gravity is proportional to mass, whereas at the centre point of the Earth the mass is zero, so gravity is zero.

"PROPOSITION IX. THEOREM IX. That the force of gravity, considered downward from the surface of the planets, decreases nearly in the proportion of the distances from their centres." (Ibid., 406)

Physics tells us that the centre of the Earth is the hottest point. It can only be the centre of tension and can't be the centre of gravity. But like gravity, tension is proportional to mass, and its strongest point is not at the centre of the Earth but at the crust. **The crust is the strongest junction of the two forces: the convex surface is the strongest gravitational boundary, and the concave surface is the strongest tension boundary at which the two forces coincide.** Going back to the new definition of gravity, the force of gravity between objects can only occur on surfaces that touch each other.

The total gravity of the Earth calculated earlier is about 660 billion tons, which is the force of gravity and tension of the Earth's surface. When leaving the Earth surface, the tension force immediately disappeared, while gravity continues to play its role with air as a medium but the intensity drops rapidly, disappearing about 10 kilometres above the surface. Therefore, objects here and higher still fall to the ground, due to the effect of pressure.

Tension is associated with high temperature, and its nature is antigravity, which causes the expansion of celestial bodies. The source of gravity is the cohesive force of particles that tend to be low in temperature, leading to the contraction of celestial bodies. Contraction multiplies the tension, while expansion does not increase gravity, contrary, it weakens it. This is one of the reasons why the tension of celestial bodies is stronger than their gravity and why a celestial body generally presents as a spherical shape. As mentioned earlier, radiation is an antigravity force. Generally speaking, the tension of the universe is stronger than that of gravitation. That is, antigravity is a dominant force. The universe is not fragmented, it is held together, mainly by whirlpool forces and shells.

The foregoing discussion can be summarized as two new propositions of the gravity-tension of spheres. First, the gravity-tension of the sphere centre is zero. Second, the strongest gravity-tension of a sphere lies on the surface of the sphere. These propositions can be called gravitation-tension hollow theory and gravitation-tension edge theory respectively. Their role is twofold: to retain gravity and to remove the problem of action at a distance.

Reference. In *The Principia*, Newton had a lot of descriptions about the gravity of the sphere. Contrary to my description here, he thought that gravity was centripetal force (*The Principia*, 3, 287), pointing to the center of the sphere. Objects outside the sphere exert weight on the sphere, that is, gravity; Objects inside the sphere and objects on the surface of the sphere are attracted to each other, but their attractions (gravity) are completely canceled out.

"Therefore the forces of these particles exerted upon the body P are equal between themselves. For the forces are as the particles directly, and the squares of the distances inversely. And these two ratios compose the ratio of equality.

The attractions therefore, being made equally towards contrary parts, destroy each other. And by a like reasoning all the attractions through the whole spherical superficies are destroyed by contrary attractions. Therefore the body P will not be any way impelled by those attractions." (Ibid., 189)

"PROPOSITION LXXI. THEOREM XXXI. The same things supposed as above, I say, that a corpuscle placed with out the spherical superficies is attracted towards the centre of the sphere with a force reciprocally proportional to the square of its distance from that centre." (Ibid., 190)

These two paragraphs illustrate two states and two outcomes of the particles. 1. 'particles exerted upon the body P', 'a corpuscle placed within that superficies will not be attracted by those forces any way.' 2. 'a corpuscle placed with out the spherical superficies is attracted towards the centre of the sphere...'

In other words, **attraction (centripetal force, gravity) just occurs between two or more isolated objects.** In the first case, these particles and body P are in fact combined into one, and mutual attraction does not exist. In the latter case, the particle is outside of P, and there is mutual attraction.

Newton didn't say here that 'gravity is proportional to mass', that body P would attract all the particles, including air molecules and dust out the surface, into that surface, thus eliminating the mutual attraction.

New Definition of Gravity

Picking up from the foregoing discussion, we can define gravity **as the polymerization force generated by mutual contact and resonance of corpuscles under pressure, which is proportional to the pressure and the number of corpuscles in contact with each other.**

In other words, gravity is the aggregation force of corpuscles (particles, atoms, and molecules), which is the aggregation force of material fields. Particles aggregate into fields and nebulae; quarks aggregate into atoms; atoms aggregate into molecules; and molecules aggregate into everything. All of these aggregations are the result of the action of gravity, which is the force generated from the mutual contact of corpuscles. The CFT highlights aggregation force, but this book retains the concept of universal gravitation only for convenience of discussion.

Particles are varieties of photon in the material world, they have independent minimum space, no positive or negative distinction, and no mutual attraction, but mutual repulsion. Their aggregation requires pressure. Starting from the contact point, include compounding, joining, hooking, blocking, clamping, friction, impinge, tapping, radiation, etc., particles that meet will aggregate with each other in some way, which is gravity. **Gravity is an interaction force. It is not a force that corpuscles are born with but is a force produced by their contact.** Any object in isolation, including black hole and mucilage, has no gravity. If, when contacted, no aggregation of corpuscles occurs, such as with a pure sand pile, then they have no gravity either.

A generally neglected force is vibration and its reinforcing form, resonance. It allows the particles to permeate each other, close contact, eliminate all cracks and pores, and even merge together. Without this property, pressure alone would produce only a weak gravitational pull between particles, and what really holds them together is resonant force.

Radiation acts as a repulsive force on objects and also excites the gravitational attraction of particles, i.e., the Casimir effect, a radiating contact.

NOTE. "Casimir effect: The attractive pressure between two flat, parallel metal plates placed very near to each other in a vacuum. The pressure is due to a reduction in the usual number of virtual particles in the space between the plates." (Hawking, 2016, p.236) I interpret it as the radiation and resonant force of particles between objects. It is not gravity, let alone virtual particles action, but particle aggregation force transformed from radiation pressure & resonant force. **Although radiation is a kind of repulsive force, an antigravity, but it excites resonance, and is therefore gravity.**

A good example of a daily activity to illustrate my point is dough mixing. When we add the right amount of water to the flour, we stir and knead it to allow the molecules and atoms of the flour to come into wide and close contact with each other and with the molecules of water. The loose flour thus becomes an elastic and tightly bound dough in the end. So, gravity comes from nothing, from weak to strong, and the pressure is converted into a cohesive force, namely, the resonant force. Baking powder strengthens their resonance and expands their volume, but still a sticky lump.

However, the pressure in this example is an external force, while the pressure on the nebular particles is mainly vortex force. It is not an external force, but a force transformed by the flight and rotational motion. This force forces a large number of free particles to lose their freedom, trapping themselves in larger particles and clumps. Flour also has a large number of free particles, and the water molecules slow them down and tend to coalesce. The pressure generated by kneading not only imprisons them, but also clumps them together. We can't knead a pile of sand into a ball because each sand grain is a solid cage of particles, which makes it impossible for the grains to touch tightly and resonate. The dough, however, will lose its elasticity if it is kept for a long time in air, due to the disappearance of external pressure, and the activation of free particles.

Objects are made up of small particles that have distinct individuality, and most of their contact is indirect, that is, through intermediaries. Celestial bodies touch each other through light fields and radiation, and objects on Earth touch each other through air, so the attraction between them is weak, even less than the repulsion.

Under the action of vortex force, air molecules extensively make contact with the Earth's surface and, through interaction, form a new layer of molecules close to the ground, which becomes a stable foothold for the

atmosphere and also a stable foothold for all objects within it. The interactions between bodies and air, and between bodies and bodies, are still carried out through the contact of their molecules and atoms.

This gives rise to a mechanical rule: **All forces come from the movement of objects and are also manifested *in* the movement of objects, which is generated from their mutual contact and resonance. An isolated object has only energy or potential energy, no any force.**

The Forces to Fix Things on the Earth Surface

When we fly high in the sky by plane, we often see the bright Sun above, a flat sea of clouds below, and an atmosphere supporting the sea of clouds farther below. Readers who have read the foregoing section may think that clouds and the upper atmosphere are completely formed by the rising gases that are lighter than the gravitational index. Why do they not continue to rise? The advantage is that the more upward (outward) one is, the thinner the air and the weaker the gravity from the Earth, and therefore the less the resistance there is to rise. So what force is stopping these rising gases?

Air consists of free molecules, free atoms, free ions and is essentially discrete, its fluidity shows that it is not bound by gravity. It can be preserved for billions of years, mainly because it has coalesced into a continuous whole, a tight shell of matter that envelops Earth and forms atmospheric pressure. On the outside, the ozone layer and the ionosphere become the shells of the atmosphere.

Under the dual action of the vortex effect & shells, the direction of the Earth spinning force changes from the outward discrete to the inward convergent, which forces the natural divergent motion of the atmosphere into horizontal flows, forming atmospheric pressure. This is the real reason that air, seawater, and other surface objects cling steadily to the earth's crust.

Among these shells, the ionosphere has the highest temperature, and its strong radiation has internal and external effects that are suppressing the atmosphere inside, and rejecting the Moon outside.

According to the tug-of-war model, the moon does not have the tendency to rendezvous with Earth, and the repulsive force of the ionosphere makes it even less likely to do so. The reason why it does not leave is the constraint of a larger shell, the combination force of Venus- Earth-Mars three orbits.

The sun is formed by the Swirling Kuiper Belt nebula. After that, its swirling shape disappears, retaining the swirling motion. **All objects in the solar system will be permanently trapped in this invisible sphere, and not even artificial detectors will be able to fly out. Similarly, ground objects cannot escape the Earth.**

I can assert that without the pressure caused by vortex & shell, all celestial bodies and objects on them will be drifting loosely in space, and colliding with each other randomly.

Factors that Constitute the Weight of Ground Objects

According to Newton's law, gravity is proportional to mass. The mass and gravity of the Earth is a definite value. It exerts the same gravity on all ground objects, and the weight of an object is entirely determined by its own mass. There is **a paradox here: not all objects with mass have weight, such as hydrogen, which not only has no weight but also can rise up, showing a negative weight.** In our understanding, this is buoyancy. Like pressure, it works on all ground objects. Like seawater, air is both pressure and buoyancy.

From Newton's third law of mechanics, we conclude that **buoyancy is counter pressure, which equals and cancels out pressure on a straight line,** but we feel the pressure everywhere, which is the weight of matter. The fact that an object has weight suggests that the pressure on the atmosphere is slightly greater than the buoyancy and is rooted in the mutual gravitational attraction between the Earth and the atmosphere, known as the gravity index, which is proportional to density.

A body must be denser than the gravitational index to produce weight. Weight is not proportional to mass but proportional to specific gravity. The figure 1.293 kg/m^3 is the standard weight grade unit, abbreviated as SWGU. If an object weighs 2.58 kilograms at one cubic meter, then it is 2 SWGU. At 129 kilograms, it's 100 SWGU, and soon.

We can do it in another way. Imagine an isolated cubic meter of space on the ground, which is 1 SWGU when it is filled with air. Double its air and double its density, and its weight is 2 SWGU. Add 100 times more air and 100 times more density to get 100 SWGU, and so on. The weights of various objects can be converted in these two ways. For example, one cubic meter of water is 1,000 kilograms, equivalent to 775 SWGU (1,000/1.29). Physicists have long worked out specific gravity tables based on this principle.

SWGU only applies to the Earth's surface. As the air pressure decreases with elevation, the SWGU decreases accordingly. In a vacuum, all objects are in a state of weightlessness; the mass of the object remains, but its weight disappears.

SWGU doesn't work at all on the moon. There is no air on the moon, so there is no atmospheric pressure and no buoyancy, but an object still has weight. Its molecules and atoms fuse with molecules and atoms on the surface of the moon, creating a gravitation that measures weight. In addition, there is also a hot shell above the moon or in the space common to the moon and earth, but no one has yet found it, which also forms pressure and body weight.

The evidence. The moon's mass is about 735 billion gigatons, which is about 1/81 of the mass of the Earth, and according to Newton's laws, that's how strong its gravity is. It's more than 13 times weaker than what physicists now determine to be the moon's gravity, which is about 1/6 as strong as Earth's; the only explanation is the unknown shell's function.

Flying objects such as airplanes and birds are out of contact with the ground, so gravity has little effect on them. They use energy to keep their buoyancy higher than pressure and fly below the SWGU. This is their secret that allows them to stay in the sky for a long time. Now, this idea has become common knowledge.

Humankind lives in the atmosphere, as fish live in water, and is subjected to the triple forces of pressure, buoyancy, and gravity. The combination of pressure and gravity is greater than buoyancy, which gives weight to people and ground objects. Fish use their body fat to reduce their density and stay in the water. The same goes for people going up in hydrogen balloons: lower the overall density of the people and vehicles below the SWGU.

Finally, this discussion results in an antigravity proposition: **every object on the surface of the Earth, if vaporized and reduced to a density below SWGU, would**

be weightless and have no gravity.

Another antigravity proposition is that **gravity between bodies entails mediation and external pressure, and without these two conditions, gravity cannot be produced.** In other words, in a vacuum, an isolated object has no gravity and no weight. In this case, stars don't attract each other, but they repel each other as a result of their strong radiations.

NOTE. Newton denied the notion 'gravity as essential & inherent to matter'.

I ascertained in 'New Definition of Gravity' that there is only corpuscular aggregation force, no gravitating force between them. Even the aggregation force is external, because each corpuscle has its own limit space, that is, Avogadro constant, which prevents them from combining, they are generally mutually exclusive. The mass of an object formed by the aggregation force is not necessarily its weight. Its weight is not caused by the combined gravity of itself and the earth, but by external pressure. **Magdeborg hemisphere shows the strong atmospheric pressure, the source of all the object weight, not the gravity of the earth, which can be judged from its direction of exertion—It's a lateral pull. Gravity in particular cannot counter the centrifugal force generated by the earth spinning, thus prevent the loss of the atmosphere.**

Without the intense pressure created by vortex, not only the air, but most ground objects, including seawater, rubble, dust, vehicle, animal and people, would be weightless and drift away from the Earth. All that was left of the ground were mountains, forests, meadows, and sturdy buildings. Objects floating in the sky do not orbit the Earth like comets and satellites, but are lost in the vast expanse of space. Before that, the moon was long gone.

Jupiter Is a Small Star That Glows Intermittently

Judging by its high hydrogen content, high spin-speed, and seventy-nine moons, **Jupiter is not a large planet but a small star that glows intermittently, planets in particular lack the ability to make moons.** Shells and high angular momentum cause Jupiter to contract repeatedly, heat up, glow, and explode, while small mass causes it to drop in density and dim after an explosion and enter into a long period of unlit rest, similar to volcanoes on Earth. Jupiter has exploded several times before, producing many moons and rings. If the current spin rate remains, Jupiter's intermittent glow cycle will continue, and the other three wood-like planets, especially Saturn, will be similar to Jupiter.

The popular view is that star luminescence is the result of thermonuclear reaction. According to Newton's law, the mechanism of thermonuclear reaction is gravity, which is proportional to the mass of the planet. Only when the minimum mass of stars reaches 8% of the Sun's mass — that is, eighty times the mass of Jupiter — can there be enough gravity to support a thermonuclear reaction. From this, it was concluded that **Jupiter would need to increase its mass by 80 times to start thermonuclear reactions and emit light before it could evolve into a star**. Jupiter can't gain that much mass, and it can't shine as a star — but it does.

As mentioned earlier, a planet is an outfall of the Sun, and everything on the planet is a polymer of plasma. The remaining plasma forms the core of the Earth and the lava is this plasma spewing from the ground. Remove the shells of the Earth, and its core is a small star, so the core of Jupiter must be a bigger star. Whether there is nuclear fusion in Jupiter's core is still controversial, but the fact that no nuclear fusion in the Earth's core is certain. **This suggests that nuclear fusion is not the cause for stars to glow, and**

that they do not need oxygen to glow.

According to CFVT, from outside to inside, stars are generally composed of three different plasma strata: gaseous plasma, liquid plasma and solid plasma, **and accordingly, their temperatures decrease step by step. Their core temperature is the lowest, and thermonuclear reaction or nuclear fusion is impossible;** Their photosphere has the highest temperature, and the light of the stars comes from it; Their sunspots and explosions (producing satellites and rings) mostly occur on the surface of its liquid layer. All of this applies to Jupiter, its satellites and rings are also its superficial explosives rather than its core explosives.

Previously, astronomers discovered a brown dwarf star in the halo of the Milky Way galaxy and named it J0104; it is made up of gas, its equivalent mass is ninety times that of Jupiter, and it is two hundred fifty times purer than the sun. J0104 has more than 99.99% hydrogen and helium content and is considered the purest and largest brown dwarf, or a 'failed star' — confirmed to be the fate of all brown dwarfs.

However, in 2017, astronomers discovered three small bodies similar to Jupiter, named TRAPPIST-1, J0523 and J0555. Among them, J0523 is slightly larger than Jupiter and the other two are slightly smaller, but they are not included in the big planets or brown dwarfs, instead they're called small stars, or red dwarfs. J0555, with a mass eighty-five times that of Jupiter and a gravity three hundred times that of Earth, is called the smallest star ever discovered because it is 40% smaller in diameter than TRAPPIST-1. (https://www.rankred.com/smallest-stars-in-the-universe-mean-radius)

Of the five small stars, J0104 has the highest hydrogen content and the largest mass; Jupiter has a diameter comparable to theirs. **The other three small stars can become 'successful stars', and it is my opinion that J0104 and Jupiter should not be considered 'failed star'.**

This exposes the shortcomings of the HR diagram. When astronomers look at a star, they are most certain of its brightness, from which they calculate the temperature, and determine its grade — that is, the position on the HR diagram — which is fixed. Then they use Newton's law, Steffen-Boltzmann law, and so on, to calculate its diameter, mass, gravity, life, and distance from Earth and other parameters. **The HR diagram acknowledges that stars change in brightness, but only get darker as they age, not brighter.** J0104 is thought to be more than 10 billion years old, and even though it might be originally a red dwarf, it has now degenerated into a brown dwarf, no longer capable of thermonuclear reaction, and no longer a star.

J0104 high purity hydrogen seems to have been neglected here, and this explanation does not apply to Jupiter. Only intermittent luminescence is the best answer, which brings J0104 and Jupiter back into the stellar family.

REFERENCE. Some people figured J0104 as "Estimated to have formed about 10 billion years ago, VLT measurements also suggest J0104 — an L type subdwarf — has a mass equivalent to 90 times that of Jupiter, making it the most massive brown dwarf found to date." (http://www.sci-news.com astronomy/record-breaking-brown-dwarf-04731 .html)

The Misunderstood Group: Jovian Stars

We can now confirm that Jupiter and the other four are small stars of the same class, they share the same vortex as their host star and were formed at about the same time. White dwarfs are their luminous phase, black dwarfs their dark phase, and their transitional phases are brown dwarfs and red dwarfs. **They have a high hydrogen content, they must have a high spin rate (to be determined) too, which will make them glow; it is simply their small mass that makes them too small to last. They are four different phases of small star, not four different stars.**

Trappist-1, J0523 and J0555 are in the luminous phase, their brightness will increase and there will be a brief surge period, the explosions will occur around this time and produce planets and rings, then return to the long period of darkness. J0104 and Jupiter are currently in the early stages of their luminous phases, and the process since then has been much the same.

Their ultimate fate is predictable. After many explosions, the main body will shrink continuously and finally become a bunch of unlighted asteroids or asteroid belts. It is also possible that the main body will disintegrate in a crushing explosion, with a meteor shower being its final glory. Finally, there is another possibility; due to the decrease of hydrogen content or insufficient angular momentum, it cannot reverse the phase change trend and return to the plasma state, thus becoming a black dwarf.

This theory refreshes the following concepts:

1. The mass limit of stars.
   This can be less than 8% of the mass of the Sun, or even less than 1% of the mass of the Sun — that is, less than the mass of Jupiter. Among them are Saturn, Uranus, Neptune, J0523 and J0555, etc. They are not wood-like planets, but Jovian stars. From the reality of the solar system, I infer that the number of

these small stars emitting light intermittently in the universe is several times more than the number of stars emitting light continuously. Thus, the family members of stars increases several times — most of them belong to the grandchildren of superstars.

2. Properties of white dwarfs.

   The white dwarf is not the remnant core of the old star after an explosion, but **a small star that is in its luminous phase.** It has approximately the same hydrogen content and density as its parent star, it emits light intermittently, and it has a relatively short life span —Its luminous cycle will get shorter and dimmer each time. Jupiter is a typical white dwarf star during its glowing period. Its density is only one-third higher than that of water, and other Jovian stars are similar. The popular notion that white dwarfs are millions of times denser than water is based on Newton's laws, which is an intention to fill the mass gap with density.

3. The nature of a black dwarf.

   A black dwarf is thought to be the form of a white dwarf star after extinction, which is very rare because of its long process. **I think it's just the phase of darkness for small stars, and that there are several more black dwarfs than white dwarfs.**

4. Sources of supernova explosions and meteor showers.

   The early explosions of small stars are so intense that they are often mistaken for supernovae. At least some of the nine supernova explosions observed in our galaxy are caused by the explosions of small stars. Another result of Jovian star explosions is a meteor shower, which happens in the case of the last mass explosion of the small star.

5. The cause of star glow.

   It is not the result of nuclear fusion caused by mass gravitation, but **the result of vortex force compressing mass particle into plasma,** as this causes a small number of high-speed ions to escape

as starlight.
6. **Plasma is the source of everything in the Solar System.**
The 79 moons of Jupiter are very different in nature and shape, but they are all formed by its burst of plasma. It follows that the Earth, the other Terrestrial planets, and everything on them — including plants, animals, and humans — were transformed by the plasma of the Sun.

CF. Sun, iron and sulfur: How life began
Trento, 17 Luglio 2017

A new study led by the University of Trento explains the biochemical mechanism that might have originated life on Earth. The article, published in *Nature Chemistry*, unveils the role of the iron-and-sulfur clusters in the probiotic evolution, solving at the same time a long-lasting paradox about the first living cells on our planet.

To survive, all organisms need energy. But to produce that energy, living creatures depend on the metabolic activity of a complex of metal ions that are coordinated to proteins: the so-called iron-sulfur clusters.
(https://pressroom.unitn.it/comunicato-stampa)

The Causes of Saturn's Low Density

Saturn has the lowest density of all the planets in the solar system. The actual number is 0.687 grams per cubic centimetre, only half the density of Jupiter (1.33 grams per cubic centimetre), which is one-third less dense than water. If we had a large enough pool of water, Saturn would float in it, and at least one-third of it would remain out of the water.

According to this book, Jupiter-like planets are small groups of stars that formed in the same nebula as the Sun, and at the same time. They are similar in property to the Sun, so why is Saturn lighter so much? The answer is its explosion: Saturn had a violent explosion recently.

Given its small mass, a significant consequence of the explosion of Jovian stars is the reverse phase transition, where a great deal of solid and liquid plasma transforms into gaseous plasma, causing the sphere to expand, the density to drop, and its light to fade.

Judging from the shape and motion of its ejecta, Jupiter mostly explodes briefly, but in a strong way, and the blast mainly forms moons. While Saturn, on the other hand, exploded in a more balanced and continuous manner, or, a continuous eruption, mainly forming rings. In terms of time, I estimate the most recently occurring explosion of Jupiter was one million years ago. It has now passed the loose & low temperature period after the explosion and has entered a new round of acceleration, encryption, and heating, ready to glow. Saturn's most recent explosion occurred much more recently, one hundred thousand years ago. Now it is at its lowest density and temperature period. It will take five hundred thousand years for this trend to reverse — the time here is all estimated.

If we believe that satellites and rings are the outbursts of these explosions, then by observing and calculating the age of these outbursts, we could estimate a relatively accurate time of these explosions. **One thing is certain: Saturn's**

most recent explosion occurred after Jupiter's, not the other way around. Today's phenomena on Jupiter, including shrinking, encryption, heating, and phase transition, will be repeated on Saturn in the future, at that time, Saturn will shrink to half its current size.

In recent years, space probes have found that Uranus, Neptune and Pluto have very low temperatures, and large areas of DIA Sea and DIA Mountain appear on the surface, showing the trend of deep phase transformation. Will they become terrestrial planets in this way? It is too early to draw this conclusion. Because we observe the strong wind on their surfaces at the same time, which is a sign of high-speed spin and intense contraction movement, resulting in temperature rise, liquid phase change and re-luminescence.

Hypothesis. Origin of Saturn's moons and rings.

Compare Jupiter & Saturn, their spin speeds are 10.3 km/s and 12.66 km/s, and their densities are 1.33 and 0.67, respectively. From these two sets of data, according to the genesis theory of tidal sloshing in this book, although Saturn spins slower, but its density is only half that of Jupiter. Therefore, when Saturn's contraction leads to spin acceleration, apart from exploding to produce a moon, the plasma on its surface is more likely to be thrown into the air and form a ring. Explosion and throwing objects will lose angular momentum, which will reduce its spin speed and expand its volume. Then, a new round of contraction, acceleration, explosion and object throwing began, forming a new moon and ring, pushing the previous moon and ring outward. After several cycles, the overlapping moon and ring system of Saturn is formed.

Tides Come from the Sloshing Effect of Seawater

That the flux and reflux of the sea arise from the actions of the sun and moon.
Isaac Newton

Newton first linked the tide phenomenon with the gravity of the sun, the Earth, and the moon. (*The Principia, 424)*. Almost all the explanations of tidal phenomena so far are attributed to the gravity of them.

However, a phenomenon mentioned in *The Principia*, though explained by Newton, still puzzled me.

"Cor. 2. Since the moon's force to move the sea is to the force of gravity as 1 to 2871400, it is evident that this force is far less than to appear sensibly in statical or hydrostatical experiments, or even in those of pendulums. It is in the tides only that this force shews itself by any sensible effect." (*The Principia*, 468)

When the tide comes in, it surges like a thousand troops on horses that go their inexorable way, we could even use it to generate electricity. How could we not measure it at all on instruments such as seismographs, level gauges, and pressure gauges? Keeping in mind that the weight of the weighing apparatus should be slightly reduced when measuring objects.

This is a sign of superstition over distance force. If we can completely eliminate this idea in our minds, we will get inspiration from the example of a waterspout. A waterspout does not 'attract' seawater directly into the sky, but it first creates a funnel-shaped low-pressure space by absorbing air and allowing seawater to 'flow' in.

How strong is the attraction between the moon and the earth? Newton integrated some British observations and provided a comprehensive data. "Wherefore, since our sea, by the force of the moon, is raised to 8 3/5 feet, the lunar fluid would be raised by the force of the earth to 93 feet." (Ibid., 471) What a tremendous power!

If the gravity of the moon and the sun can lift the seawater on Earth, even raise the ground of the Earth and the moon as well, the air above the water surface and ground will be lifted first, thus creating a huge waterspout and leading it to land. This kind of gravity can't produce waterspouts on land, and it can certainly produce super-strong tornadoes or dust storms. Neither of these phenomena can be seen, so can we believe that gravity works solely on seawater?

Furthermore, we now know that **the ocean accounts for 71% of the earth's surface area, and if the whole ocean is lifted up by moon's gravity, which means that the whole earth is pulled up to;** The moon is a solid, not a fluid, but it also has gravity in proportion to its mass, and both of them are suspended in the air. The inevitable result of their combined gravity is that the distance between them is getting shorter and shorter, and they should have collided together long ago —If everyone had gravity, could you imagine that the skins of two people hanging in the air bulge towards each other by gravity, and they keep still?

All these facts show that the tide phenomenon has nothing to do with external gravity and is caused by Earth's own forces. A simple and clear explanation is the inertia force generated by Earth's spinning — that is, centrifugal force. The speed of the Earth at the equator is 466 m/s, faster than the speed of sound. Couple that with the revolution speed of nearly 30 kilometres per second in the same direction, it is the inertial force of surface object movement that is used by the spacecraft launched by all countries.

The combination of the two forces cannot make the sea water fly to the sky, but is strong enough to make the waves roll back and forth on the surface of the earth. "It is caused by the backward and forward motion of the waters; compare water sloshing back and forth in a basin, going up on one side as it goes down on the other." (Bacon, 2018, BOOK II, 36) Although this passage comes from Bacon's book, it is not his unique view, but the prevailing European view of the time. Newton must have been the opposition to it, because it was the antithesis of gravity.

Seawater is on the crust, which is solid, and its movement is synchronized with the whole. While seawater is a liquid, its movement lags behind, with the upper layer lagging behind the lower layer more severely. Its movement must be out of sync with the whole, and its layers are also out of sync. Seawater is like the water in several big potholes on a big ball, shaking with the spinning of the earth, touching the edge of the continent, and being blocked from flowing back, showing regular ebb and flow movements.

Even when excluding the gravitational pull of the Sun and the Moon, tidal phenomena will still occur. The force will not weaken at all. We can believe that if the Earth spins faster, the tidal movement will increase and appear as a huge wave. With further acceleration, seawater will flow back and forth between land and oceans, and even the Qinghai–Tibet Plateau will be washed back and forth at high tide. If it accelerates again, reaches the first cosmic velocity, 7.9km/s, or Saturn's spin velocity, 10.3km/s, the seawater will fly up into the air in the same direction the Earth spinning, and most of the ground objects — including beasts and humans — will also be thrown into the sky, to form rings like Saturn's.

In fact, the Earth's current spin and revolution forces are powerful enough to throw water and Earth's surface objects up into the sky. However, this phenomenon does not occur, not because of the effect of gravity but whirlpool force, expressed as atmospheric pressure, which has already been discussed.

Newton acknowledged that centrifugal force of the earth was powerful, but it was a force that raised the water level at the equator, not a tidal force. It seems that these are two unrelated movements rather than the same one. Compared with this centrifugal force, the sun's gravity is weak,

(negligible, but in the tidal force, it is one of the two major forces that can not be ignored.)

"Cor. Since the centrifugal force of the parts of the earth, arising from the earth's diurnal motion, which is to the force of gravity as 1 to 289, raises the waters under the equator to a height exceeding that under the poles by 85472 Paris feet, as above, in Prop. XIX., **the force of the sun, which we have now shewed to be to the force of gravity as 1 to 12868200, and therefore is to that centrifugal force as 289 to 12868200, or as 1 to 44527, will be able to raise the waters in the places directly** under and directly opposed to the sun to a height exceeding that in the places which arc 90 degrees removed from the sun only by one Paris foot and 113.033 inches; for this measure is to the measure of 85472 feet as 1 to 44527." (*The Principia*, 465)

At this point, the difference between me and Newton on the origin of the tides is clear. **I think the centrifugal force of the Earth, e.g., the sloshing effect here, is the main cause, whereas Newton believed that the gravitational pull of the sun and moon was the driving force of the tides, and that the centrifugal force of the Earth, no matter how powerful, was not.**

From the current map of the Pacific Ocean drawn by the geological department, we can clearly see that the sea water is not static, and the sea water on the equator flows from west to east following the spin of the earth, generating eddy currents on both sides. When the sea water in the northern hemisphere hits the edge of the American continent, it will turn back to north and east to form an ellipse. Its whole sloshing period is estimated to be more than ten hours, which is undoubtedly the root of tide.

It is also a kind of antigravity, which is more than a million times stronger than the combined gravity of the moon and the sun—Note. "Cor. Since the centrifugal force of the parts of the earth, arising from the earth's diurnal motion, which is to the force of gravity as 1 to 289"

(Ibid.465), "Cor. 2. Since the moon's force to move the sea is to the force of gravity as 1 to 2871400...Cor. 3. Because the force of the moon to move the sea is to the like force of the sun as 4.4815 to 1" (Ibid.468). Their combined gravity can not overwhelm the antigravity produced by the spin of the Earth, thus show great power on the Earth.

We can simply equate wisdom with discrimination and selectivity. If there are multiple answers to a question, I do not follow the custom but to choose the simplest, most clear, or most convincing one, and then only refer to or discard the rest. This is one expression of wisdom. Now, I heard a deafening voice:

"Rule Two. We should attend only to those objects of which our minds seem capable of having certain and indubitable cognition.

All knowledge is certain and evident cognition. Someone who has doubts about many things is no wiser than one who has never given them a thought; indeed, he appears less wise if he has formed a false opinion about any of them. Hence it is better never to study at all than to occupy ourselves with objects which are so difficult that we are unable to distinguish what is true from what is false, and are forced to take the doubtful as certain; for in such matters the risk of diminishing our knowledge is greater than our hope of increasing it. So, in accordance with this Rule, we reject all such merely probable cognition and resolve to believe only what is perfectly known and incapable of being doubted...Now the conclusion we should draw from these considerations is not that arithmetic and geometry are the only sciences worth studying, but rather that in seeking the right path of truth we ought to concern ourselves only with objects which admit of as much certainty as the demonstrations of arithmetic and geometry." (Descartes, 1985, p.12-13)

A New Astronomical Law Arising from the Slow Spin of the Moon

Recent astronomical observations have shown that the moon is at least the same age as, if not older than, the Earth, which provides the basis for this book's theory of Earth–Moon homology. They are both the products of the Sun's explosion and were close neighbour at the beginning, circling each other.

A common physical phenomenon is that when objects of different sizes fall onto a rotating surface, small objects roll fast and far, while large objects roll slowly and nearer. This is what happened to the Earth and the Moon in the early days. Many scientists now believe that the Moon's spin, which was much faster than Earth's, was slowed down by a powerful braking force. Most people think that this power is tidal force, which not only makes the Moon's spin-speed slow down but also causes the Earth's spin-speed to slow down — the distance between the two being long. Tidal force is also the source of Earth's volcano and earthquake disasters.

The source of tidal force is gravity, which is often exaggerated. This is the result of ignoring vortex force and antigravity. It is also an exogenous theory. **Any change in things has internal and external causes; we cannot ignore either side, but we should first pay attention to internal causes.** Internal factors are more dominant factors, and external factors are more counselling factors. The original spin-speed of Mercury and Venus, for example, was also very fast in their beginning, but then they slowed down. We can't attribute this slowing to the tidal forces of their moons because they don't have any moons. Jovian stars, on the other hand, have so many moons that tidal forces don't stop them spinning.

It's an astronomical fact that both the Moon and Earth are spinning slower, but the Moon is slowing down faster. We can even establish a new astronomical law stating that

all stars, planets, and moons slow down or stop spinning during their lifetimes mainly because of their own explosions, vibrations, jets, and radiation, the braking forces. Explosions create separations, such as planets and moons, that further reduce the spin velocities. New-born stars are often luminous, rapidly spinning, exploding, and energetic. While older stars tend to be dim objects with slow but steady spins. They are like the figures of the young and the old in the world respectively. But we don't have to worry about the universe becoming aged. Vortex motion and black holes continue to create new stars, so the universe is constantly being updated.

The appearance of the Moon shows that it was originally a lava body like the early Earth. Because of its small size, **under the action of vortex force, internal matter and energy were squeezed out to form craters all over the lunar surface and the spin-speed dropped rapidly, causing it to enter into a stable state.**

A planet that is similar to our Moon, is **Venus.** It has thousands of volcanoes all over the surface and due to its greater mass, its volcanoes are still active today. Its surface is more than 80% volcanic terrain and its mass is 81.5% of that of the Earth's, while its atmosphere is 93 times the mass of the earth's atmosphere and 96.5% is carbon dioxide — all of which are the consequences of volcanic activity. **The most significant consequence is the braking effect, which has no effect on its revolution but obviously impacts its spinning motion — almost interrupted.**

The slow decline of the Earth's spin-speed is an irreversible trend, and the main factors are volcanic eruptions and earthquakes. Tides and water circulation, on the ground and in space, also contribute to it.

Some people may query this as, according to the foregoing statement, compared to the Moon and the other three Earth-like planets, the Earth's geological activities are obviously the most frequent and intense, and its spinning speed should not be so fast — ranking as the fastest among the Earth-like planets. My explanation is that **the Earth**

has been shrinking. One of the vortex effects mentioned earlier is the contraction of the main body, which in turn accelerates the spin of the main body. That is why Earth has maintained a relatively high spin-speed up to now. The most obvious consequence of the Earth's contraction is volcanic eruption — magma is squeezed out by the shrinking crust, just like we squeeze toothpaste. Another, less obvious, consequence is earthquakes, caused by the deformation or fracture of a block under shrinking. Earthquakes have less effect on the speed of the earth's spin, **volcanic eruptions are expansionary and anti-contraction movements, which are the main 'braking' forces**— You can compare it to the braking mechanism of a car wheel. However, according to Newton's third law of motion, the anticontraction force is equal to the contraction force. More than 70% of the Earth's surface is covered by seawater, which absorbs or suppresses most of the energy from volcanic eruptions, allowing the contraction force to be slightly greater than the anticontraction force, thus maintaining the Earth's high spin-speed. Mercury, Venus and the Moon lack surface water, and the volcanic braking effect is strong, which is the first factor for their slow spin-speed.

In fact, there is no obvious contraction movement of the earth, which is more appropriately called pressure rather than contraction force in form. It was created by the ionosphere and planetary orbits, that is, the earth's dynamic shells. The radiation from the shells passes through the atmosphere, transforming the spin force of the earth into the surface pressure, causing the earth to shrink slightly, and the reaction force then eliminates it.

The downward trend of the Earth's spin speed is certain, but instead of being flat, it fluctuates, determined by the result of **the confrontation between the forces of contraction and expansion.** Even with the fluctuates, it has surprised us greatly with its stability and accuracy. On January 5, 2021, The Daily Telegraph of UK reported that Earth has reached its quickest spin speeds in the past half-century — the day on July 19, 2020, was 1.4602 milliseconds shorter than 24 hours, suggests that the period was dominated by contraction movement. This is the period we are living through, marked by active volcanic activities.

Inferred from several ice ages in geological history that the earth has gone through periods of expansion. Back then, the Earth spin was slower, earthquakes and volcanism were rare, it was quiet, and snow and ice were everywhere. It's supposed to coincide with the contraction, which means we're going to have a new ice age in the future? The record-breaking heatwave impacts in 2021 denied this speculation. It shows a continuous warming trend instead, a sign of the Earth contracting. According to Newton's third law of motion, however, this trend won't continue forever. It may now be nearing its peak, and the reversal is expected to come in the near future.

References from Google. Com indicating that the earth is still contracting.

1. 'This is climate change': July was world's hottest month on record By Euronews with AP. 14/08/2021
2. Climate crisis: world is at its hottest for at least 12,000 years. By Damian Carrington, Environment editor. 27/01/2021
3. Record-breaking June 2021 heatwave impacts U.S. West. By Tom Di Liberto. 23/06/2021
4. The Volcano Eruption in Tonga Was a Once-in-a-millennium Event, By Alice Klein, *New Scientist*, 17/06/2022

Comparison of Rotational Motion and Vortex Motion

There is no void. Space is full of matter and energy, including dark matter & dark energy. Therefore, no object can be always moving in a straight line; most of the time it's a curve path. What we usually see is rotational motion and vortex motion. They are curvilinear motion around a centre or an axis, of which may be fixed or unfixed, temporary or variable.

In daily life, we see more rotary motions, such as the rotation of various wheels. Most of them are solid, have a definite mass, and their energy comes from outside sources. **Its centre is determined, both the centre of mass & energy, and the centrifugal force will be thrown out of the matter inside in a kind of divergent movement, resulting in the object's disintegration. This kind of rotating objects are artificial fine products, rarely seen in nature, usually known as mechanical rotation motion.**

Vortex motion is the most common motion in nature. The galaxies that astronomers observe are vortexes of all shapes and sizes, which produces a continuous centripetal contraction force that makes the body of the rotating object smaller. At the same time, it twists the surrounding substances in the process of motion, thus making the rotating object larger. The mass and energy of the body multiply continuously in this movement, which is a significant difference from mechanical rotation motion.

Vortex motion indispensable condition is dense space and shell (boundary). Under these constraints, matter cannot diverge outward and is forced to change its direction of motion. Centrifugal force changes to centripetal force, and rotational motion changes to spiral motion.

In reality, **a marshmallow machine** is a good example. When the syrup is thrown out of the small holes of the container, it acts as a centrifugal force. If there is no obstruction, the syrup will spread around. Instead of spreading out, the syrup gathers in the centre to form an ellipsoid, which is the function of a circular cover, and it is the shell of this small vortex. It transforms the centripetal force of it into centripetal force. Careful observation will also see that the molecule of syrup is not a linear accumulation motion but a curvilinear spiral motion. This is how stars are formed from swirling nebulae.

 Hotline public picture

The main driving force of the mechanical rotational movement is mainly in the centre, and matter-energy radiates outward from the centre, which requires a continuous consumption of external energy. Once the energy is exhausted, the movement will stop immediately.

The driving force of vortex movement comes from the aggregation of particles. It increases with the increasing number of particles and a gradual tightening of power from the outside in, like the golden hoop on Sun Wukong's (the Monkey King's) head. **As the matter is continuously compressed, the aggregation energy is continuously generated and enhanced in the movement, which will accelerate the internal rotation. The core of the vortex thus expands and becomes denser at the same time, forming a tighter and hotter ball, with a higher and higher spin-speed.** This is the secret of vortex motion that enables the material nebula transform into a black hole, then, undergo an energy-mass conversion, and finally become a radiant star.

Methods to Judge the Spin Motion Properties of Stars

Stars all have spin motions. The method to determine which of the foregoing two motions a star exhibits is as follows.

If it is a mechanical rotating motion, it must be a divergent motion. No matter how big or small the star, matter is constantly thrown out into space, and the spheres become smaller and smaller, eventually breaking up into a ring, an asteroid group or belt. The higher the star spin-speed, the faster the process.

If it is a vortex movement, because it is a closed and restrained motion, matter and energy will not be lost and the process will be much more complicated, with the following four phenomena occurring successively:

1. **Thermal contraction and increase in spin- speed**
   A continuous process of contraction in which the sphere heats up, rather than cools down (this case occurs in vortex) due to the continuous contraction. It completely violates the law of heat expansion and cold contraction, which is its most prominent feature. The contraction of the sphere means that matter and energy of it are compressed, the tension is increased, and the spin is accelerated

2. **Phase change**
   At this time, the vortex force can no longer make the sphere double its rate of contraction. Instead, it becomes pressure to increase the density of the sphere (the star), causing it to undergo **a phase change from gaseous state, to liquid state, to solid state.** There is a low temperature in the core of every sphere, so the phase change must start from the surface and go in deep. It can never reach the centre or develop into a whole solid state, it can only form a gradually thickened liquid layer mixed with more or less pure solid substances, such as

diamonds. Most stars are in a high-temperature state and cannot undergo phase change. Only small stars will undergo distinct phase change during their nonluminous period; at present, Jupiter is in this state. Uranus, Neptune and Pluto have even higher phase transitions, and scientists have found massive diamond seas, diamond showers, and iceberg-sized diamonds on them. But as it heats up under rotational pressure, the liquid layer and diamond sea will revert to a plasma state again and the phase transition will change direction.

The Earth and several other Earth-like planets were also gaseous spheres when they were born. Due to their low hydrogen content, high nitrogen, other element content, and low temperature, phase transition persists and no reverse transformation occurs. Instead, their crusts are preserved, and organic life and human beings evolve among them.

3. **explosion**

   The result of continuous contraction movement and phase change will not weaken the vortex force. On the contrary, the vortex force will become stronger, because of the contraction in terms of volume. This will push the sphere to further encrypt, heating and glowing. The high temperature of the core is not conducive to heat conduction. Layers of shells cut off the heat way out, causing sphere explosion. Most of the explosions are local small-scale explosions rather than large-scale explosions, and the explosive materials form low-temperature spheres such as planets and satellites. All stars have undergone a series of explosions.

4. **Evolving into a black hole**

   Stars with large mass and high spin-speed will turn into black holes, then ignite, and explode into new celestial bodies.

Since shells and vortex motions are common, I can definitely say that **most celestial bodies exhibit vortex motion and are controlled by vortex forces. Celestial bodies that exhibit mechanical rotation are few and limited to cold, small bodies.** For example, the asteroids between Jupiter and Mars are all non-spherical bodies, and most have no spin motion; they either remain the debris of small star explosions or collide with each other.

Evolution of the Earth

The earth, though in comparison of heaven so small, nor glistering, may of solid good contain more plenty than the Sun, that barren shines.
John Milton: *Paradise Lost*

The Jupiter-like planets and the Sun are made of the same vortex stuffs, they're the same age and they have the same simple compositions and luminescence. Their structure is similar to that of the Sun, with gaseous plasma on the surface, liquid plasma in the middle, and solid plasma at the core. The difference is that they are small in mass and do not emit light continuously, only intermittently. They will glow and explode in the future, which is why this book has christened them with the new name of Jovian stars.

Earth-like planets are the outbursts of the early Sun; they're younger than sun and made up of gaseous and liquid plasma. Due to the small mass, continuous contraction — under the action of vortex force — suppresses the free ion movement, thus most of the gaseous ions phase into liquid ions (lava) and form a solid core. In a whirlpool motion and at early high temperatures, simple hydrogen ions begin to fuse and interact to produce more diverse and complex elements. This is the process in which ions are trapped in atomic shells, and molecular and object shells gradually lose their freedom. This is the process in which Earth-like planets evolve from plasma to atom-molecular bodies, shifting from a relatively violent state to a relatively calm and stable state, and the temperature drops again and again. Of these planets, the Earth is blessed with the most perfect evolution.

Vortex forces continue to work, causing Earth to shrink and compress, resulting in a phase change on the surface from gaseous, to liquid and solid, with rock layers and oceans appearing. Several glacial periods in geological history are also a part of this phase change. Shrinkage causes the spin-speed of Earth to accelerate, which then cyclically increases the rate at which the body is shrinking. As a result, the solid and liquid components further increase, while the gaseous components decrease and remain on the surface. The hydrogen ions then coalesced into more than 100 kinds of atom.

The reaction force with force, goes hand in hand: the stronger the force, the stronger the reaction force. This triggers volcanoes, earthquakes, and thermal radiation, that slows the contraction and spin-speed of the Earth, making its motion balanced and stable. This then promotes the combination of hydrogen, oxygen, carbon, sulphur, and other elements, to form underground mineral deposits, surface organisms, and human beings, which is Earth as we know it today.

Action against reaction is universal. It causes all stars and planets, not just the earth, to continue to contract, and it also causes the opposite movement. **The contraction results in an increase of heat and spin acceleration, while the anti-contraction results in cooling down and spin deceleration.** With the two movements alternating, I see it as the basis for the sunspot cycle, the Jovian stars intermittently luminous cycle, and the earth glacial pulsation.

Solar System Q & A

Q. Why does the author believe that planets are outbursts of the sun and not exotics?

A. **Most of the sun angular momentum is distributed to the planets.**

In *One, Two, Three... Infinity: Facts & Speculations of Science*, George Gamow wrote about the origin of our Globe as well as to the origin of every other planet of our solar system:

"A few decades later entirely different views concerning the origin of our planetary system were formulated by the famous German philosopher Immanuel Kant, who was more inclined to think **that the sun made up its planetary system all by itself without the intervention of any other celestial body.** Kant visualized the early state of the sun as a giant, comparatively cool, mass of gas occupying the entire volume of the present planetary system, and rotating slowly around its axis. The steady cooling of the sphere through radiation into the surrounding empty space must have led to its gradual contraction and to the corresponding increase of its rotational speed. The increasing centrifugal force resulting from such rotation must have led to the progressive flattening of the gaseous body of the primitive sun, and resulted in the ejection of a series of gaseous rings along its extended equator... The rings formed in this way were supposed to have broken up later and to have condensed into various planets circling at different distances around the sun.

These views were later adopted and developed by the famous French mathematician Pierre-Simon, Marquis de Laplace, who presented them to the public in his book *Exposition du systems du monde*, published in 1796...

When such a mathematical treatment was first attempted sixty years later by the English physicist Clerk Maxwell, the cosmogonical views of Kant and Laplace ran into a wall of apparently insurmountable contradiction. It was, in fact, shown **that if the material concentrated at present in various planets of the solar system was distributed uniformly through the entire space now occupied by it, the distribution of matter would have been so thin that the forces of gravity would have been absolutely unable to collect it into separate planets. Thus the rings thrown out from the contracting sun would forever remain rings like the ring of Saturn,** which is known to be formed by innumerable small particles running on circular orbits around this planet and showing no tendency toward 'coagulation' into one solid satellite.

The only escape from this difficulty would consist in the assumption that the primordial envelope of the sun contained much more matter (at least 100 times as much) than we now find in the planets, and that most of this matter fell on the sun, leaving only about 1 per cent to form planetary bodies.

Such an assumption would lead, however, to another no less serious contradiction. Indeed if so much material, which must originally have rotated with the same speed as **the planets do, had fallen on the sun, it would inevitably have communicated to it an angular velocity 5000 times larger than that which it actually has. If this were the case, the sun would spin at a rate of 7 revolutions per hour instead of at 1 revolution in approximately 4 weeks."** (p.301-302)

"These considerations seemed to spell death to the Kant-Laplace views, and with the eyes of astronomers turning hopefully elsewhere, Buffons collision theory was brought back to life by the works of the American scientists...found itself treading on muddy ground." (p.303)

With reference to Kant and Gamow, my view on planetary birth is this: **'the sun made up its planetary system all by itself without the intervention of any other celestial body.' The newborn sun, a blue giant, spun wildly, it had 'an angular velocity 5000 times larger than that which it actually has.' In shells, its angular momentum was**

converted into centripetal contraction force, resulting in its series of explosions, and the exhalation formed the planets. At the same time, most of its angular momentum was transferred to the planets, which became the impetus of their spin and revolution. The spinning velocity of the sun thus dropped to its current state. Similar events occur on planets, resulting in a large number of satellites and slower spinning.

Q. Astronomical observation shows that Mars has experienced a series of intense eruptions and produced a mountain as high as 27,000 meters. If volcanic eruption acts as a brake, why is its spin velocity still so fast?

A. **Braking only occurs in opposing eruptions,** which happens from time to time on earth. From craters or volcanoes evenly distributed across the surface, we can infer that such events were once frequent on the moon and Venus, causing them to spin very slowly. **Things are very different on Mars, its eruption site is fixed. Without opposing eruption, there will be no braking effect, and even its spin may be accelerated from this unidirectional eruption.**

Reference. "We estimate that the volcanic ash observed in Arabia Terra is the result of between 1,000 and 2,000 individual explosive eruptions over 500-million years. Our observations support the hypothesis that Arabia Terra hosted supereruptions in the late Noachian-early Hesperian that repeatedly blanketed the region with layers of ash."

('Stratigraphic Evidence for Early Martian Explosive Volcanism in Arabia Terra.' By Patrick Whelley, Alexandra Matiella Novak, Jacob Richardson, Jacob Bleacher, Kelsey Mach, Reagan N. Smith. First published: 16 July 2021 https://doi.org/10.1029/2021GL094109)

Q. Why is it that the main cause of climate fluctuations is the earth contract-expansion cycle rather than man-made?
A. The primordial Earth was a hot plasma. If no contraction-expansion cycle occurs, it will continue to cool, and the temperature will be the lowest now, which even lower in the future. Geologists reject this trend, pointing out that the earth experienced five ice ages long before our time and temperatures were much lower than today's. The only plausible explanation is that **earth contraction-expansion cycles dominate earth climate.** Man appeared in the late glacial period, indicating that the appearance of glaciers on the earth had nothing to do with man, and then the temperature rise could not be man-made. Now the global warming may be related to human activities, but it must be secondary.

REFERENCE. "In the very beginning of earth's history, this planet was a giant, red hot, roiling, boiling sea of molten rock—a magma ocean...Scientists have recorded five significant ice ages throughout the Earth's history: the Huronian (2.4-2.1 billion years ago), Cryogenian (850-635 million years ago), Andean-Saharan (460-430 mya), Karoo (360-260 mya) and Quaternary (2.6 mya-present). Approximately a dozen major glaciations have occurred over the past 1 million years, the largest of which peaked 650,000 years ago and lasted for 50,000 years. The most recent glaciation period, often known simply as the Ice Age, reached peak conditions some 18,000 years ago before giving way to the interglacial Holocene epoch 11,700 years ago." Extremescience.com

Q. What is the cause of Cambrian Explosion?
A. About two billion years after birth, the Earth went through a major phase transition, turning its magma (plasma) ocean into water ocean under the force of the vortex contraction. As sea water evaporated, a dense atmosphere formed. It blocked out sunlight and pushed the Earth into the first ice age, a long period of cold darkness. (Reminder.

the devastating eruption of Mount Tambora in 1815 created a Year Without a Summer of 1816 in the world.) The Earth continued to contract, leading to frequent volcanic eruptions and earthquakes that caused temperatures to rise and land to surface. After that, the earth undergone a contraction-expansion cycles, with a corresponding major glaciation period-minor glaciation period cycles. We are now at the end of the most recent minor ice age.

The second ice age was critical to life. Later in the duration, atmospheric density dropped, temperatures rose, and most glaciers melted, the Cambrian explosion took place as vast quantities of air molecules and atoms fused with water and earth to form complex objects and living things.

REFERENCE. "Cambrian explosion, the unparalleled emergence of organisms between 541 million and approximately 530 million years ago at the beginning of the Cambrian Period. The event was characterized by the appearance of many of the major phyla (between 20 and 35) that make up modern animal life." Encyclopedia Britannica.

Q. Why don't the great deserts fall in the relatively cold and dry north, but in the relatively warm and wet Africa, central Asia, south America and Australia?
A. The popular explanation is human destruction of forests. That's not a good answer, because most of the areas that have been deforested by man have not become deserts. From the current ecology of Australia, I dub the title of the first desert builder to herbivores.

Rabbit and sheep are docile animals, which are not invasive and can only eat grass ceaselessly. However, without intervention, look, in less than a century, they will turn the remaining green lands in Australia into deserts. This is how deserts are created all over the world.

A large number of archaeological findings confirm that today's desert areas were originally the paradise of plants, and also the blessed land of man and animals. After men had killed & domesticated most beasts, the enemies of herbivores were removed, and they, both wild and domestic, could reproduce freely, which greatly exceeded the growth of plants, leading to the arrival of the great famine. **Finally it came to this day, even the tree roots and glass roots were grazed up by them— the key to turning fertile soil into desert, —and they were covered in deserts made by themselves at last.** Whales, manatees, otters, hippopotamus and other herbivores changed to foraging in the water in time to survive.

Beasts feed mainly on herbivores, though fierce and cruel, they save the forests and hold a pretty world of diversity.

Q. What is an unexpected benefit of human activities proposed by the author?
A. Men dig underground mineral deposits, build railways, highways and big cities, operate all sorts of transportation vehicles and machines, wage wars and conduct nuclear weapon tests. **All these make the world restless but make the earth a little more peaceful, reducing big earthquakes to small ones. A few big earthquakes are driven into the sea and underdeveloped areas.**

Our generation has witnessed this miracle. Large earthquakes occurred frequently in the last century, but they are rare in the 21st century, and most of them become small and medium ones. China's southwest areas are identified as a seismically active zone and have been quiet for nearly half a century, the result of intense activity by men and machines. It is expected that there will be fewer and fewer earthquakes of magnitude 5 or higher in inland of the world in future.

The principle is clear: **A devastating earthquake requires a long period of energy accumulation and the activities of men and machine can accidentally interrupt it and trigger it prematurely.**

Reference: 'Big City Defuses Big Earthquake', by Solatle Lu. *Imagination Science*, 2010.

Q. Why does the author elevate man to the supreme position in the universe?

A. All creative activities are guided by the **active spirit. Nature has no active spirit. All her movements and products, however magnificent and delicate, are results of necessity**—may be a grand creation, the supreme good; may be a great destruction, extremely evil in our eyes.

Nature has created a variety of plants and animals in our planet, but the fishing ape is the best of them, who relying on his head and hands in the struggle to evolve into human beings. Man is the only creature with an active spirit, and naturally becomes the highest being in the universe.

Man's most pretentious merit is to create a spiritual cosmos, to become the soul of the material universe. The origin universe thus obtains its full life.

In theory, there could be other advanced intelligent life in the universe. But they cannot surpass humans in intelligence, for we already possess the highest knowledge and the ultimate truth of the universe. Man deserves to be the soul of the universe. The most optimistic estimate is that the aliens are equal to us, or slightly superior to us technologically, if they could visit our planet.

The Most Common Products of Vortex Motion: Spheroids

Anyone with a basic knowledge of astronomy knows that a good portion of things in space are shaped like a sphere. Stars, planets, and moons are all spherical, and it all comes down to gravity. All the atoms in an object pull toward a common centre of gravity, and they're resisted outward by whatever force is holding them apart. The final result could be a sphere.

At the beginning of the 21st century, an international team in Hawaii, USA, made an accurate measurement of the Sun by means of the transit of Mercury and found that the Sun's equatorial bulge was only 12 kilometers or so, almost a full circle. (https://science.sciencemag.org/content/337/6102/ 1638.abs tract)

The spherical phenomenon is common but ascribing it to gravity is hardly convincing. Just imagine that in space there is a mass of sand equal to the mass of the Sun (one step farther than the nebula) and that gravity alone would automatically coalesce it into a ball and turn it into a glowing sun? I bet this will never happen.

One objection is enough: antigravity is everywhere, and it is stronger than gravity. Tug-of-war is an antigravity phenomenon that prevents grains of sand from forming clumps. Gravity also creates another force, tension, within its own products that is opposed to itself. Tension increases at a greater rate than gravity because it is the explosive force that is trapped inside an object by its outer shell.

Only vortex motion can create celestial bodies, which has been discussed previously. As for why its products are all round balls, the answer is clear.

No matter what the original shape is, the core of the vortex movement is a circle ball. Everything around it is rotating in a centripetal motion. The product formed is

naturally a circular ball. Vortex motion also produces tension, which is the uniform force of expansion that causes an object to become round, like we inflate a balloon. Stars, no matter how big or small, are the products of swirling motion, so they are round and spherical in shape. The same is true of larger and smaller objects.

We know from our daily production activities that most products in rotary motion are circular. For example, in porcelain processing, no matter what shape the raw materials are, they immediately turn around after being put on a ceramic rotating machine for several minutes. The pots, jars, bowls, cups, and such that are produced in this way are naturally round. This principle is equally valid in the sky and is even more effective. So, while the ceramic processing surface is an incomplete three-dimensional space, close to the plane, **stars are produced in a totally enclosed circular space that is a perfect three-dimensional space. The material from the circular inner wall is uniformly concentrated toward the core, and the energy from the circular inner wall is uniformly applied to the core. Thus, the perfect sphere is naturally produced.** Regarding the sand in question, if it was placed in a nebula and the hidden energy of each grain of sand was released, it would spin and eventually form a glowing sun.

From astronomical observation results, we can come to the conclusion that **all round and oval celestial bodies are the products of vortex motion. The irregular celestial bodies including comets and asteroids are mostly fragments generated after small stars explode or collide.**

NOTE. In reality, most celestial bodies are not perfectly round, but have a **certain flatness**. That's because their spin throws materials, in particular gaseous and liquid stuffs, towards their equator. The higher their spin speed, the greater their flattening, which is already a physical common sense.

Why Scientists Ignore the Greatest Forces in the Universe

The good may be said to be not only the author of knowledge to all things known, but of their being and essence, and yet the good is not essence, but far exceeds essence in dignity and power. Plato: *The Republic*, 509e

The greatest force in the universe is not electromagnetic force, nor nuclear force, nor gravitational force, but vortex force generated by vortex motion. The ubiquitous vortex galaxy in space proves this truth. Why do scientists turn a blind eye and a deaf ear to the fact? **Like micron explorers sitting rapt at the center of a tornado, focusing on forces inside grits, and disregarding to the swirling force that is lifting them up into the sky.**

First, it needs an eye of the soul. Plato believed that the eye of the soul is far more precious than ten thousand bodily eyes, for by it alone is truth seen. (*The Republic,* 527e) In fact, it is an ability to recognize and grasp the essence of things; Plato called it the supreme goodness, which is wisdom. This ability, generally speaking, means philosophers are superior to scientists, but it is rare even among philosophers. "Inasmuch as philosophers only are able to grasp the eternal and unchangeable, and those who wander in the region of the many and variable are not philosophers." (Ibid.,484a)

Second, the vortex theory is imperfect. The vortex movement of the universe is a function of cosmic force. **Both Leucippus' and Descartes' vortex theories lacked the concept of the cosmic force, and due to poor observation they are unable to see the celestial vortexes clearly and get the eternal facts of vortex motion,** resulting in their theories being unconvincing and easily refuted.

Third, corpuscularianism overrides holism. Vortex theory belongs to holism, and atomism tends toward corpuscularianism. Democritus inherited both of these two theories from Leucippus but emphasized atomism more, which was adopted by later generations. **Atomism and particle theory thus became the arterial road of science, walked down by many, while vortex theory and holism were abandoned together.**

Fourth, Newton was firmly opposed to vortex theory. Vortex force is a trillion times stronger than gravity, but **Newton replaced vortex theory with the theory of gravitation, and his theory was strongly supported by Kant and Einstein.**

Fifth, there are queries about and misunderstandings of the spiral motion of galaxies. **After the Greeks, no one took vortex phenomena seriously except Descartes, and there was no systematic theory of celestial vortexes before CFVT,** so doubts and misunderstandings became common, three of which are highlighted in the following sections.

Query One: Why Can't We See the Winding?

"It takes less than a 100 million years for the random initial structure to evolve into a spiral. **The milky Way, however, is 100 times as old. This means that the original structure must formed many more spirals. Like the grooves of a long-playing record, the spiral should have been coiled the center 100 times and more. Yet we have found no such thing.** The spiral arms or a galaxy, as seen in plate ll are not tightly wound, and hence they cannot be the remnant of an embryonic structure. Since none of the observed spiral systems shows very close spirals, we must assume that the spirals do not wind around." (Kippenhahn, 1993, p.221)

When we stir a cup of tea or coffee, we create a vortex. This vortex has three distinct features: microgrooves, heavy objects (tea leaves) at the centre, and bulges at the edges, the first of which astronomers doubt the most.

In this regard, I think the first issue is attitude: opposition or approval. If you are an opponent of CFT, you will not see the centrally located dense print at all. Even if you occasionally see it, you will give irrelevant explanations. If you are a believer, you will follow the trend of increasing the centripetal curvature of the spiral arm and believe that the central microgroove will be produced and that the reason for the rare microgroove can be found.

The reason for this is also common: the galactic vortex is dominated by stars and nebulae, which, when rolled in large numbers to the tapering head of spiral arm and the edge of a black hole, are violently torn apart and merge into sheets or clumps known as **quasar.** If there are multiple spiral arms, they will also converge there to form larger quasar. The motion of a quasar is very violent and irregular — regular microgrooves cannot be formed. Even if there are a few microgrooves on its edge, it will be covered up or spoiled by the light of the quasar.

In addition, given the large number of stars, the spiral motion of galaxies cannot be completed in one go. For example, after one hundred million stars fall into black holes, energy-mass conversion takes place and new celestial bodies are created, and such events will continue. Big bubbles and giant stars observed by astronomers in the galaxy core are the products of this conversion. Under the action of vortex force, they also explode and destroy the dense grains.

However, as long as we broaden our vision, we will see that **all the spiral arms are curved. Some of them have to make one or more turns before entering the centre, and these are macrogrooves in a conspicuous swirling motion.**

Hotline public picture

Query Two: Why Do We Not See the Shortening and Disappearance of Spiral Arms?

This question is — based on the fact that the big bang has lasted for more than fourteen billion years and the Milky Way is extremely old — why are its spiral arms still so long? Here, we might as well regard the rotating arm as a tornado moving in the opposite direction. The common points of the two include that they are conical in shape and make spiral movements, which can roll into the surrounding objects and expand themselves. The difference is that the tornado's suction port is at the small head on ground, and its contents are pushed up and ejected to the wide tail. It relies on external energy sources, mainly solar energy, and its life is short.

The power of the spiral arm is the aggregation energy of particles, which is never exhausted and will be enhanced with the increase of inhalants. The speed of movement is originally high, but the following two factors make it slow down and prolong its service life.

Firstly, absorption is onerous. It has a large tail of suction, mainly stars and nebulae, and is rich in resources that will not stop absorbing until it has cleaned up its surroundings and cleared out its isolation space.

And secondly, the transport task is arduous and full of resistance. The huge amount of spiral arm material is pushing towards the relatively small head, and the space is getting smaller and narrower, meaning the chances of rubbing and colliding with each other are increasing, and the speed of travel is getting slower and slower. That's why we don't see any shortening of the spiral arms.

Available observations show that the four spiral arms of the Milky Way show no signs of shortening, but rather of extending, and that a fifth spiral arm is also in the process of formation. It is quite likely that by the time the Solar System has spiralled into the galactic core black hole in the next 10 billion years, the five galactic spiral arms will still be extending. Even if human beings can live to that day, they may not be able to see the spiral arms shortening or disappearing.

According to an article titled *Black hole sports huge 'bubbles'* contributed by Davide Castelvecchi in *Nature* magazine (Heywood, I. et al. Nature573, 235-237-2019), **astronomers have spotted giant 'bubbles' around the Milky Way black hole;** it's the first major result from South Africa's pioneering MeerKAT radio telescope. This bubble is rich in heavy elements, and some think it's the remnant of a supernova energetic explosion at the galaxy centre nearby about 100,000 to 50,000 years ago. Others such as

Q. Daniel Wang of the University of Massachusetts, Amherst, think it's vomit from the galactic black hole. He argues that colossal black hole at the heart of the Milky Way is a messy eater. Of all the gas that falls toward the black hole, 99% gets spewed back out into space and that forms this bubble. I follow the latter view. With temperatures of millions of degrees Celsius and speeds of 1,000 kilometers per second, it looks more like the eruption of a black hole than the remnants of a supernova explosion tens of thousands of years ago.

Astronomers have also found that black holes in other galaxies eject a lot of materials. Like the Milky Way, these materials will definitely enter the spiral arms and form new stars, which will become the subsequent resources for the growth of the spiral arms.

Black holes don't simply swallow and spit out stars. They refine and renovate stars like a smelting furnace, making the universe forever young and alive.

In G.Gamow's time, the black hole had not been discovered yet, he had the genius to guess that the spiral arms were the ejecta from the center of the vortex, and the spiral arms are the breeding grounds for new stellar populations. His argument supports my view above.

"In comparing the theoretical calculations of Jeans with Hubbles empirical classification of galaxies, we find that these giant stellar societies follow exactly the course of evolution described by the theory. In particular we find that the most elongated shape of elliptic nebulae is that corresponding to the radius-ratio of 7/10 (E7), and that it is the first case in which we notice a sharp equatorial edge. **The spirals that develop in the later stages of evolution are apparently formed from the material ejected by the rapid rotation,** although up to the present we do not have a completely satisfactory explanation of why and how these spiral forms are formed and what causes the difference between the simple and the barred spirals." (Gamow,1947, p.292)—They have one and two core black holes respectively.

"Much is still to be learned from further study of the structure, motion, and stellar content in the different parts of galactic societies of stars. A very interesting result was, for example, obtained a couple of years ago by a Mt. Wilson astronomer, W. Baade, who was able to show that, whereas the central bodies (nuclei) of spiral nebulae are formed by the same type of stars as the spherical and elliptic galaxies, the arms themselves show a rather different type of stellar population. This 'spiral-arm' type of stellar population differs from the population of the central region by the presence of very hot and bright stars, the so-called 'Blue Giants,' which are absent in the central regions as well as in the spherical and elliptic galaxies. Since, as we shall see later (Chapter XI), **the Blue Giants most probably represent the most recently formed stars, it is reasonable to assume that the spiral arms are so to speak the breeding grounds for new stellar populations. One could imagine that a large part of the material ejected from the equatorial bulge of a contracting elliptic galaxy is formed by primordial gases that come out into the cold inter-galactic space and condense into the separate large lumps of matter, which through subsequent contraction become very hot and very bright."** (Ibid, p.293)

Query Three: A Lot of Stars Have Been Observed Moving Faster Than the Spiral Arms

There seems to be evidence that stars are not inherently part of the spiral arm and must leave it.

My answer is as follows: vortex motion is common, not only in galaxies but also in systems higher and lower than the galaxy, so while **the spiral arm participates in the galaxy vortex motion, it also makes smaller scale vortex motion itself, rotating forward like a band of cannonballs — this is the reason why stars cannot get away from their spiral arms.** The stars at the front travel at two speeds at the same time; that is, the rotation speed of the stars and revolution speed of the spiral arm, and they appear to us to be moving faster than the spiral arm itself. Astronomers might also see some stars moving slower and in the opposite direction from the spiral arms, which is what happens when stars pass on the other side of the galaxy spiral arm, similar to what we can see from a Chinese revolving lantern, a merry-go- round, a stunt cycling show, and a tornado.

To extend this principle to super-galaxies, astronomers' assumption that the Milky Way and Andromeda will collide in three billion years might be also a visual illusion. The two galaxies belong to the same local galaxy group, and what the astronomers observed was only their opposition movement on different rotating arms or in different sides of the same spiral arm, and the collision was not bound to happen.

If in the latter case this collision is successfully avoided, the opposition movement may occur again or more times, but the periods will be shorter and shorter, and they will finally really hit each other, and together fall into the galactic black hole.

Scientists are used to applying solar system models to the Milky Way, in which all objects revolve around the core of the galaxy, just as the planets revolve around the Sun. If you calculate on this basis, the Solar System orbits the Milky Way at about 250 kilometers per second, with a period of about 220 million years, and currently it has made about 20 revolutions.

This book suggests a more complex overlap model. That is, **stars — in addition to spinning themselves — experience three kinds of motion. The first is the rotation around the axis of the spiral arm, the second is the precession towards the galactic centre under the vortex contraction pressure, and the third is the revolving motion of the spiral arm to the galactic center.** Among them, the velocity of the second motion determines the life duration of the solar system.

A New Theory of Black Hole Genesis and Evolution

Big Bang cosmology holds that the giant star formed a black hole by collapsing under its own powerful gravity after the thermonuclear reaction ended. This black hole is a low-temperature object that has radiation. The radiation of small black holes is strong, and their life span is short. Large black holes radiate less and live longer than the universe.

CFC holds that all **celestial vortexes have a centre with high-temperature, high pressure, and high spin-speed that accelerates matter to the speed of light, known as black holes.** Spiral arms continuously push celestial bodies inward, making them stronger. Thus, as matter accumulates, the energy-mass conversion mechanism is activated, and the black hole ignites light, becoming a new plasma body. **A black hole is the foetal state of a new star, a transitional body.** It disappears with the birth of the new body. Therefore, its life span is always shorter than that of the star from which it comes, just as the survival period of the foetus is shorter than that of the human beings from which it has evolved.

Again, I want to mention the role of the spiral shell. **The numerous celestial bodies and massive nebulae surrounding a black hole in a galaxy are the black hole shells, as are the spiral arms and quasars, which is as important to the formation and evolution of a black hole as the womb is to a foetus.**

Quasar plays the most unusual role in these shells. It is both the food of the black hole and its most direct shell — in the period before being involved in the black hole, its strong light radiation effectively blocks the leakage of matter and energy from the black hole, including the black hole radiation, which ensures the existence and growth of the black hole.

The **'ring conjecture'** put forward in 1972 by Kip S. Thorne, professor of Feynman Physics at California Institute of Technology, is a kind of shell theory. He imagines that an object, such as a star or a cluster of stars, undergoes highly aspherical compression and only when its circumference in all directions is less than the critical circumference, will it form a black hole surrounding it. In *Black Hole and Time Warps: Einstein's Outrageous Legacy*, he mentions that he had been persevering with this topic for a quarter of a century, during which time Cornell University's Stuart Shapiro and others simulated his conjecture on supercomputers and failed to get a definite result.

In my opinion, Thorne's 'ring conjecture' should be confirmed. The reason why his research subject miscarried lies in the basic theory. It is not the collapse of an old giant star that forms a black hole, but the vortex that makes it. In addition, most black holes and their shells resemble spheres rather than rings or loops.

How Vortexes Make Black Holes Easily

From the latest observations and the law of spheroidal contraction, we can determine an astronomical fact: at the core of all celestial vortex motion, including stellar vortex motion and galaxy vortex motion, there must be one or two black holes.

The role of the shell cannot be ignored. The curved arms are also shells for the core —temporary shells. Their restraints on the vortex include the following:
1. **The vortex movement is kept in a closed state from beginning to end. Matter & energy can only flow in and not out.**
2. The vortex ball must be **contracting all the time;** its ultimate goal is a point, a singularity.
3. **Density & spin velocity increase simultaneously. To abide by Scaling Law,** all spheres are perfect three-dimensional balls. For every time the vortex ball shrinks by double its volume, its (mass & energy) density & spin velocity will increase simultaneously by eight (2^3) times. On this basis, the second doubled contraction will result in the increasing of its density & velocity of the vortex ball sixty-four times of the initial, and so on.

Based on this, the calculation formula can be established:

DCV = $(2^3)^n V_0 = 8^n V_0$

Here, DCV is the velocity generated after the body doubled contraction; n, the numbers of doubled contraction; V_0, and the initial velocity of the body can be m/s, km/s, etc.

Scientists love the example of figure skating: **when the figure skater opens her arms, her spin slows down, and when she draws in her arms, she goes faster.** This describes the inverse relationship between the volume of the spinning body and the spin-velocity of the same body, showing that the energy accumulation of the spinning body causes autogenous compactness and accelerated rotation. The example is true for vortexes, which are unidirectional contractile motions. Within vortexes the energy is constantly compressed, and the results are beyond imagination.

Hotline public picture

The skater's volume changes by less than 10% depending on how far her limbs are drawn in, and a doubled contraction is impossible. But the nebula can do this easily because from the nebula to the star, the main body has shrunk by hundreds of millions of times; the Sun is formed by the contraction of a nebula the size of the interior of the Kuiper belt sphere. In fact, at a tenth of its doubled shrinkage, the velocity of matter has been accelerated to 1073741824 times (8^{10}) its original velocity. **Let's say the initial velocity is one meter per second; now it is 1073741824 meters per second, which is more than three times the speed of light. Under the action of this internal rotation force, the substances press against each other, and the pressure continues to increase. Their molecular shells, atomic shells, and nuclear shells are therefore destroyed in turn, and the particles are completely released and all convert into energy, form a black hole.**

According to Daniel Bernoulli's principle, the black hole becomes the centre of low pressure because the spin-velocity increases again and again, and the matter outside continuously flows into it, becoming energy. **The light beam then disappears because light beams and matter are inseparable.**

This corrects a popular view that the gravitational field of a black hole is so strong that it only allows foreign matter to enter but not any internal matter to flow out — even light can't escape. Hawking had confirmed black hole radiation and new astronomical observations have confirmed that black holes, which not only radiate and absorb material at all times but also periodically eject massive amounts of material outward, **are the most active bodies.**

The Black Hole: The Smelting Furnace of Celestial Bodies

The previous discussion paints a new picture of a black hole, not as a gravitational body that even light can't escape from, but as a furnace that matter comes in and out of. The spiral arms of galaxies are not what some astronomers call illusory patterns but are real streams of matter. **The spiral arm acts like a mechanical conveyor belt, sending a steady stream of stars into the vortex's centre. There, they are torn apart into strips and chunks. That's what astronomers call quasars, which are then completely crushed and spun into black hole, transforming into indistinguishable energy. Afterward, as external matter increases, temperature and pressure continue to soar, activating the energy-mass conversion mechanism, in which energy is converted into the simplest substance: hydrogen ions, that form new substances and new celestial bodies.**

Not all the stars on the spiral arms are old stars. On the contrary, they are mostly young and middle-aged stars. They come from their own black holes, and they follow the spiral arms to galaxy black hole. Whether young or old, they are devoured by the bigger black hole, contributing to the casting of new celestial bodies.

Stellar black holes, stars and some of the bigger planets are formerly black holes, after the birth of a new star, no longer exist; it is itself the new star. The black hole is only its transitional form, the fetal state, which is different from a smelting furnace. Black holes also occur in high-level and low-level celestial systems; Super-galaxies, and even the universe, go through this process but it is impossible for humans to see these scenes.

All galaxies and super-galaxies have large black holes at their cores, and their evolution is far more complex and lengthier. For obvious reasons, they need to deal with huge amounts of matter, and despite all their power of digestion, black holes have a limit. Once that limit is reached, say 8 million solar masses, its energy-mass conversion mechanism kicks in and new objects are created. Instead of disappearing, the black hole continues to absorb matter and build stars.

In recent years, this sort of event has been observed many times by astronomers in the United States and Europe, with the most prominent observation being the large galaxy containing black holes in the centre of the **constellation of Phoenix.** They are surrounded by millions of degrees of hot gas and these gases are ejected from black holes. The mass is equivalent to a few mega-solars, which is several times larger than the mass of all the galaxies in the Phoenix galaxy. They are the materials of new stars and some new stars have been formed.

Note how this description differs from the popular saying. **The popular view is that black holes rely on strong gravity 'to attract' objects in, and keep them, including light, can no longer escape. While I think that the black hole is the lowest pressure center of the vortex, and the surrounding objects automatically 'flow' in and are 'spiralled' in by the spiral arm, and will become new objects coming out of the black hole.**

We know from astronomical observations that vortexes are hierarchical and nested. The upper vortex is made up of the lower vortexes and has greater scale, higher orbital speed and a longer life span. Its black hole also has these characteristics, and thus ends up absorbing all of its lower vortexes. But we can't imagine a total cosmic vortex, because the universe is infinite and unbounded. All vortexes have boundaries — shells. Therefore, our imagined cosmic vortex, no matter how vast, cannot contain the total universe; that is, the infinite and unbound.

Quasars Are the Outer Shells of Spiral Black Holes in Galaxies

Stars are born in spiral arms, and their common destiny is to be swept into the center of the galaxy, but they don't immediately darken and fall into the black hole. Instead, they shine with unprecedented intensity, I call it the star's most brilliant pre-death glow.

In that time, they are ripped into brighter bars, and their mass & energy are spread out rapidly in space, forming the bright shell of the black hole, known as its material event horizon. I identified them as quasars five years ago. **To be sure, the cores of all quasars are black holes. Wherever we find quasars, we can find a black hole within.** However, we can't say that quasars must exist where black holes exist, because black holes have a longer life span than quasars.

For this and others reasons, **quasars exhibit great energy and redshift values.** Based on the high redshift value, most astronomers believe that quasars are old objects, born in the early universe and now located at the edge of the universe, and that they are accelerating their escape velocity all the time. This is one of the bases of Big Bang cosmology & the cosmic expansion theory, and it is obviously a prejudice that would be overturned by future observations.

CFVT holds that every galaxy has one or two black holes at its centre, and quasars are bright boundaries created by black holes & spiral arms, which are as ubiquitous in space as galaxies. Because their brightness is determined by the number of stars involved in the black hole horizon, they exhibit great instability and are short-lived. As for the super velocity of light and high redshift value of quasars, it is not difficult to understand this phenomenon. Simply imagine the stars as spectral bands that are suddenly pulled apart when they reach the end of the spiral arm and the edge of a black hole. The latest and most powerful example is the **Asassn-19bt event,** captured by TESS

(Transiting Exoplanet Survey Satellite), which was posted on the internet. Scientists call this kind of phenomenon a tidal destruction event (TDE), (https://www.nasa.gov/feature/goddard/ 2019/nasa-s-tess-mission-spots-its-1st-star -shredding-black-hole) while I call it a spiral destruction event, SDE.

In April 2019, the first photo of a black hole taken by human beings was released on the internet, (https://www.sciencenewsforstudents.org/article/black-hole-first-photo- event-horizon-telescope) showing that the black hole is not completely dark as it has a bright shell. Most people think of this shell as the accretion disk of a black hole. I am probably the only one who explicitly declares it to be a quasar, which is the inevitable result of vortex motion; the insatiable black hole cannot exist without the enormous supply of spiral arms, namely, quasars. Some astronomers have long speculated over the connection between quasars and black holes. Hawking's book mentioned, "We also have some evidence that there is a much larger black hole, with a mass of about a hundred thousand times that of the sun, at the center of our galaxy. Stars in the galaxy that come too near this black hole will be torn apart by the difference in the gravitational forces on their near and far sides. Their remains and gas that is thrown off other stars, will fall toward the black hole...It is thought that **similar but even larger black holes, with masses of about a hundred million times the mass of the Sun, occur at the centre of quasars.**" (2016, p.110); but no one explicitly proposed this as a theory because it is incompatible with standard cosmology.

The first black hole photographed is located in the centre of the M87 galaxy, which is exactly what Hawking mentioned in his book as an example. Astronomers claim its mass is 6.5 billion times that of the Sun, and such a huge mass can only be provided by the rotating arms. There can be no other source.

In May 2019, astronomers from the University of California were shocked to spot, from the WM Keck

Observatory, that Sagittarius A* was actually emitting bright flares of energy and was rapidly glowing 75 times brighter than normal for brief periods — the time interval was more than 2.5 hours.

They said the burst was brighter than anything ever produced by this particular black hole. I would like to add here that **this is the closest and most definite quasar observed by human beings so far.** But the black hole at the centre of the Milky Way is not as quiet and inactive as many astronomers think. What they observed was a galactic black hole gobbling up a huge number of stars, a common event in the universe that happens all the time.

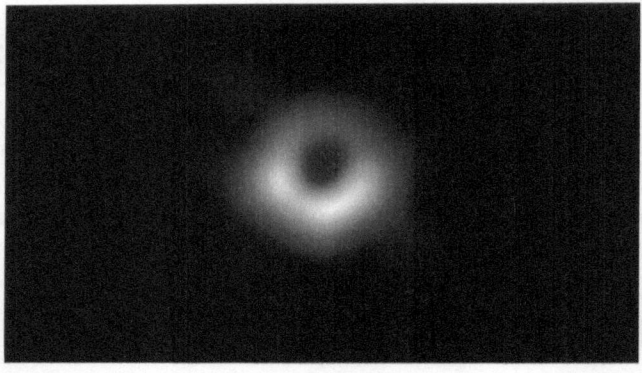

The first photo of a black hole-Hotline public picture

Black Hole Pulses Were Mistaken for Neutron Stars

"A neutron star is thought to be a cold star, supported by the exclusion principle of repulsion between neutrons."
(Hawking, 2016, p.240)

"Landau pointed out that there was another possible final state for a star, also with a limiting mass of about one or two times the mass of the sun but much smaller even than a white dwarf. These stars would be supported by the exclusion principle repulsion between neutrons and protons, rather than between electrons. They were therefore called neutron stars. They would have a radius of only ten miles or so and a density of hundreds of millions of tons per cubic inch." (Ibid.p.95)

"Thomas Gold...pointed out that **a neutron star was small enough and dense enough to be able to rotate about its axis in four seconds or less.** What's more, it had already been theorized that a neutron star would have an enormously intense magnetic field, with magnetic poles that need not be at the pole of rotation. Electrons would be held so tightly by the neutron star's gravity that they could emerge only at the magnetic poles. As they were thrown off, they would lose energy, in the form of radio waves. This would mean that there would be a steady sheaf of radio waves emerging from two opposite points on the neutron star's surface." (Asimov, 1993)

I think this kind of star, from theory to reality, does not exist.

Reason one, neutron stars have not been observed. Contemporary observation methods are very advanced. A gravitational wave that occurred 1.3 billion years ago and lasted less than one second was discovered. **If thousands of neutron stars had existed, they would have been found.**

Reason two, collapse theory fails. A neutron star is

thought to be a giant star that collapses in seconds to a neutron mass only twenty to thirty kilometres in diameter, which is physically impossible. Because **star shrinkage is a process in which matter and energy rapidly converge, at the same time — densification, heating, and explosions will follow, interrupting the process.** That's what Newton's third law of motion does.

"Assuming the 'collapse theory' of novae and supernovae, we must first of all ask ourselves about the causes that could lead to such a rapid contraction of the entire stellar body...Any attempt to contract faster than that would immediately result in the liberation of additional gravitational energy, which would increase the temperature and gas pressure in the interior and slow down the contraction." (Gamow,1947, p.324)

Reason three, the low temperature. All stars are plasma, with a high temperature and no charge. An indispensable condition for a star to evolve into a neutron star is low temperature, it needs to be cooled before plasma can fuse into atoms. **It continues to cool down until it is close to absolute zero, to allow all the matter in it to become superconductors.** Then the atom loses its dynamic and charge, and its electrons fall into the nucleon and may collapse into neutron stars. This process, as discussed earlier, cannot happen. Moreover, **astronomers have observed the opposite: the temperature of 'neutron stars' is much higher than the temperature of the Sun's surface, it can get up to 10 million degrees Celsius. This high temperature — certainly from quasars — will only reduce the neutron state of the material to an ionic state, and it will not be an opposite process.**

Reason four, the discrete effects of brittleness and high spin-speed. The spin-speed of a neutron star is very high, reaching dozens of hundreds of revolutions per second, which is a powerful centrifugal force. The cooling also makes the neutron star brittle. **If it can be generated, it will**

soon be thrown into a large area. In addition, it should be noted that neutrons have no charge and that neutron stars cannot overcome the repulsive force of the incompatibility principle between neutrons.

Reason five, the lack of independence.

"When physicists strip neutrons from atomic nuclei, put them in a bottle, then count how many remain there after some time, they infer that neutrons radioactively decay in 14 minutes and 39 seconds, on average. But when other physicists generate beams of neutrons and tally the emerging protons — the particles that free neutrons decay into — they peg the average neutron lifetime at around 14 minutes and 48 seconds." (Natalie Wolchover. 'Neutron Lifetime Puzzle Deepens, but No Dark Matter Seen', Quanta Magazine, February 13, 2018)

A neutron is formed by a proton absorbing an electron and forming an atom with the proton. When the proton disassociates from the atom and becomes **a free neutron, it has a lifetime of just over ten minutes,** and then it releases an electron and a neutrino, and reverts itself to proton, a fate that a neutron star cannot avoid.

However, denying neutron stars requires answering a statistical query: where do the billions of high-frequency pulses come from? My answer is clear: it is from black holes, or the event horizons of black holes — their quasars. When the stars are whirled to the edge of the black hole, they are instantly torn apart, shining for billions of miles and radiating outward. As the quasar passively spins along the black hole at high speed, its radiation pulsates regularly, becoming a high-frequency pulse. **Nine times out of ten, the high frequency pulsing that astronomers now receive from 'neutron stars' comes from quasars and their terminators — vortex black holes** — including galactic and stellar black holes.

"The Crab Nebula pulsar was the first optical pulsar

discovered—the first visible neutron star.

The Crab Nebula pulsar released X rays, too. About 5 percent of all the X rays from the Crab Nebula emerged from that tiny flickering light. The connection between X rays and neutron stars, which seemed extinguished in 1964, thus came triumphantly back to life." (Asimov, 1993)

From this description, we can see that **scientists have always regarded pulsars and X rays as neutron stars,** and now I have corrected them to be the products of the black hole boundaries of galaxies, namely quasars.

A Supernova Explosion Is a Phenomenon of Galaxy Renewal

When the spiraling of a galaxy reaches its end, all spiral arms disappear, the quasar phenomenon disappears, and the protogalaxy as a whole becomes a pure black hole. Under the action of continuous vortex force, the black hole contracts, leading to a continuous increase in pressure, density, and temperature, initiating the energy-matter conversion mechanism, and the black hole turns into an colossal luminescent celestial body, a nova. Then, **the nova explodes, resulting in what astronomers call a supernova, which is the forerunner of binary stars, star clusters, or a new galaxy.** The reason most stars appear as binaries is that supernova explosions often split in two, at least as a result of the first explosion.

As mentioned earlier, the spiral motion of galaxies cannot be completed in one go, and the core explosion will continue, which is a more common supernova phenomenon. Astronomers have recorded nine supernova explosions in the Milky Way galaxy that belong to this type of galaxy vortex stage explosion, or superstar explosion, not the final explosion of the galaxy. The most recent explosion occurred relatively late, making the galactic centre less active than other galactic centre, like the centre of the Andromeda galaxy — the previous discovery of Sagittarius A* burst of brightness could overturn that assertion.

It is completely wrong to regard this phenomenon as a flashback before the death of an old superstar. Even if collapse occurs, the old celestial body cannot have such a huge amount of energy. Some astronomers also observed that blue stars are present before and after supernova explosions, and Andromeda NGC205 is a striking example.

The galaxy was discovered by French astronomer Charles Messier in 1773 and named M110. It contains about

10 billion stars and is thought to be a dying or dead dwarf elliptical galaxy. Since the end of last century, it has been observed to have a few blue stars and a large amount of bright interstellar matter around its centre, which has the raw materials to form new stars. In fact, it has a tendency to form new stars, which puzzled astronomers greatly.

Rudolf Kippenhahn, a contemporary German astronomer, noted in his *100 Billion Suns* (1993) that the supernova number 1987A was originally a blue star. But the expected neutron star was not found, which puzzled him. This is consistent with his confusion about the vortex phenomenon. All astronomers who observe the vortex phenomenon and do not believe in vortex theory inevitably experience such confusion.

The mechanism of black hole metabolism allows galactic vortexes to prolong their lives, but they are not eternal. The reason is that a larger, more advanced system on top of them will have a higher energy-mass conversion threshold and will eventually evolve into the black hole equivalent of a trillion solar masses. The day it ignited, glowed and exploded, was the birth of a new cosmos.

Vacuum Is a Virtual Space, Physical Space Is Warped

Space as a whole concept is unbounded; the universe as a whole, is a space. **As long as there is an atom in the universe, a vacuum will not exist and there will be no void.** Taking particles, light balls, and fields into account, the universe is completely filled up with no void at all.

But I hold on to the idea of vacuum, because physics needs it. A fundamental proposition of physics is that time and space are conditions for the motion of matter, and this 'space' is assumed to be a vacuum. Without this consensus, calculating the motion and distance of matter would be difficult or impossible. Field theory also requires vacuum conditions.

I am assuming that vacuum is the background of the physical universe, overlapping with physical space, in which everything in the universe exists and moves. It's just that vacuum doesn't interact at all with any matter or energy. It's completely virtual, and only exists as a condition when we construct a physical model. Due to this characteristic of the vacuum, the total matter and energy of the universe can be preserved without loss, and the various motions of objects can proceed unimpeded, except when encountering obstacles in the physical world itself.

Confusing vacuum with physical space and identifying a vacuum as some kind of space of particles and energy, is one of the factors that causes confusion in physics. At first, the idea was that a vacuum is space void of matter. It leaves a place for energy. Physicists couldn't get an absolute vacuum in the lab, so they modified this concept to such, suggesting that a vacuum is a region with a gaseous pressure much less than atmospheric pressure, and quantified it further with 'Torr' as the unit of calculation. Atmospheric pressure, 760. Low vacuum, 760 to 25; Medium vacuum,

25 to 1×10^{-3}; High vacuum, 1×10^{-3} to 1×10^{-9}, etc. Then they introduced fields into the vacuum, that became a boundless energy space, which hides the infinite vacuum energy. **Some people even go far, they think the vacuum is the source of everything.**

"Mass–Energy Equivalence Extension onto a Superfluid Quantum Vacuum
Amrit Srečko Šorli
Abstract

In contemporary physics, the model of space–time as the fundamental arena of the universe is replaced by some authors with the superfluid quantum vacuum. In a vacuum, time is not a fourth dimension of space, it is merely the duration of the physical changes, i.e. **motion in a vacuum. Mass–energy equivalence has its origin in the variable density of the vacuum.** Inertial mass and gravitationalmass are equal and both originate in the vacuum fluctuations from intergalactic space towards stellar objects." (Sci Rep. 2019, Aug 13)

Today, many people see vacuum energy as a fifth force, or dark energy, in addition to the four forces of nature, and some countries are ambitious about exploiting this infinite, free energy.

The universe, as vast and boundless as it is, has only one space, the physical space in which we exist, and vacuum can only be virtual, not real. In this physical space, vortex motion is the most universal and therefore curved. But it's not space itself that's curved, it's the motion of everything, from particles to galaxies. **In other words, it is the vortex motion of everything that causes the space to be curved, rather than the curved space that forces everything to make vortex motion. In addition, shells are everywhere in physical space, constraining all motion and making them all curved, hence curved space. such as the orbits of the planets. This makes it completely different from Einstein's theory of curved space.**

INTRODUCTION: My view of spacetime, matter & energy

1. **Space and time are one and inseparable.** In the cosmos, there is no space without time, and there is no time without space.

2. **Trinity of things-time-space.** In the cosmos, there is no inanimate spacetime—there is no vacuum and vacuum energy.

3. **The independence of spacetime.** Everything, from particle to galaxy, has its own non-transferable spacetime. All spacetime come into being with things, die with them, and are consistent with the motions and changes of them. The spacetime that everything lasts is its life-span.

4. **The hierarchy of spacetime.** The spacetime of objects overlaps and crisscross each other, forming a broader spacetime. In other words, the higher system accommodates the lower system, the wider spacetime accommodates the narrower spacetime, and the cosmos accommodates everything.

5. **Energy does not occupy spacetime.** Energy is light. It either fly freely and interact with things; Or hidden in things, and composed of them, no self independent inviolable spacetime.

6. **Spacetime is an artificial concept.** "It is true that the comparison would not be worth much, for a living being is observable, whilst the whole of the universe is constructed or reconstructed by thought." (H. Bergson, *Creative Evolution*)

Also, everything and its spacetime is constructed or reconstructed by thought. Spacetime is neither matter nor energy, so it has no physical properties and laws of motion. The so-called arrow of time, flat space, curved space, etc., all refer to the motion of matter. Motions of matter are irreversible, "What is done cannot be undone." — Spacetime reversal is impossible. For the same reason, there are no objects more than three dimensions (length, width, and height) in the cosmos, so there is no more than three-dimensional space plus a time dimension, viz., Einstein's four-dimensional space.

Design a Dynamic Star Diagram

The Hertzsprung–Russell diagram (HR diagram) is a scatter plot of stars showing the relationship between the stars absolute magnitudes or luminosities, and their stellar classifications or effective temperatures. More simply, it plots each star on a graph, displaying a star's brightness against its temperature (color).

The key point of this diagram is that everything that appears to distinguish one star from another — temperature, luminosity, size, life span — is determined almost entirely by one factor: the mass of the star. The main sequence along the HR diagram is not a singular evolutionary path but a portrait of the sky at one moment in time with stars of varying masses. I call this a static star diagram. The new one is a dynamic star diagram.

The key element in the dynamic star diagram is the star's nature, the hydrogen content. New-born stars are high in hydrogen, and their light is blue and bright. But the opposite of high temperature (blue) tends to be low density, so the blue star doesn't have to be a giant. The hydrogen content of middle-aged and old stars decreases, and accordingly the star's color becomes yellow, and its brightness is lower. But when the quantity of its helium and heavy metals goes up, so does its density, so it's not necessarily a small star. **The complete process of star life is as follows:**

Irregular Nebula → Vortex Nebula → Elliptic Body → Red Giant →Black Hole → Blue Giant → Main Sequence Star + Dwarfs + Planets → Quasar → Entering Galaxy Vortex Black Hole → The components of a Supernova.

Under the action of vortex force, they keep shrinking, and the reaction force causes them to explode continuously, creating numerous small bodies, namely dwarfs, planets, satellites, etc.

All stars go through the process of evolution from type O to type M. The early Sun was a type O star that is now in its middle age. The life of a star is determined not by its mass but by its distance from the central black hole of the galaxy. This book replaces the theory of nuclear fusion luminescence of stars with the theory of plasma luminescence; the life of a plasma is infinite.

It has long been envisaged that star maps should reflect not only the state of the sky, but also the evolution of stars. American astronomer Henry N. Russell, one of the authors of HR diagram, had this idea in mind, and more people now think this way and are doing research along these lines, however my basis is completely different from theirs.

Particle Physics: Fireworks Theory

"They frequently examine difficult problems in a very disorderly manner, behaving in my view as if they were **trying to get from the bottom to the top of a building at one bound,** spurning or failing to notice the stairs designed for that purpose. **Astrologers all do likewise: they do not know the nature of the heavens** and do not even make any accurate observations of celestial motions, yet they expect to be able to delineate the effects of these motions. So too do most of those who study mechanics apart from physics and, without any proper plan, construct new instruments for producing motion." (Descartes, 1985, p.20-21)

"Therefore a separation and solution of bodies must be effected, not by fire indeed, but by reasoning and true induction, with experiments to aid; and by a comparison with other bodies, and a reduction to simple natures and their forms, which meet and mix in the compound. In a word, we must pass from Vulcan to Minerva if we intend to bring to light the true textures and configurations of bodies on which all the occult and, as they are called, specific properties and virtues in things depend, and from which, too, the rule of every powerful alteration and transformation is derived." (Bacon 2018, BOOK II,7)

Cosmologists and physicists intend, through high temperature and the violent collision of the particles, to find new particles and exotic particles to achieve the unification of particles, and to solve the mystery of the universe. This approach is similar to that of gunpowder experts who intend to achieve the unification of sparks by setting off fireworks and searching for new and strange sparks, and who then seek to solve the mystery of gunpowder.

The former sees hundreds of 'new' particles, whereas gunpowder experts see hundreds of 'strange' sparks. If they were to pay more attention, **they may also find numerous**

pairs of sparks (positive and negative pairs of particles) fighting, dancing, flirting or entangling! This is a reductionist path, full of twists, mists, and pits. The failure of this approach is inevitable. But this is the way Westerners have been walking and they will not give up easily.

The main tool is the particle accelerator. After American physicist and Nobel Laureate Ernest Lawrence built the first cyclotron in 1930, there was a boom in building such machines. Each new one increased in performance, size, and, it seemed, success. However, in my opinion, **most of the results are misleading, illusory, and failing to guide scientific research.** Hawking was certainly a reductionist, and he had a good idea about particle research. In his book, he wrote the following:

"The Weinberg–Salam theory exhibits a property known as spontaneous symmetry breaking. This means that what appear to be a number of completely different particles at low energies are in fact found to be all the same type of particle, only indifferent states. **At high energies all these particles behave similarly. The effect is rather like the behaviour of a roulette ball on a roulette wheel.** At high energies (when the wheel is spun quickly) the ball behaves in essentially only one way—it rolls round and round. But as the wheel slows, the energy of the ball decreases, and eventually the ball drops into one of the thirty-seven slots in the wheel. In other words, at low energies there are thirty-seven different states in which the ball can exist. **If, for some reason, we could only observe the ball at low energies, we would then think that there were thirty-seven different types of ball!"** (2016, p.136)

If anyone builds a billiard collider and compares the spin-speeds and directions of the balls before and after collision, they will also find many 'strange' phenomena that are not usually seen, thus obtaining a large number of research results. Such a researcher can also repeat the experiment and get more and more achievements. Finally, however, he doesn't know what billiard balls are. Whether that's the case for particle physicists? I think they are very akin.

Here we go back to a topic discussed earlier in this book: holism or corpuscularianism.

One of the major achievement's humankind has made on the road to exploring truth is having built a foundation for various disciplines. Physics is based on atoms and motion; chemistry, on elements; biology, on cells; medicine, on organs; psychology, on the nervous system; evolution, on the human spirit; astronomy, on stars; cosmology, on human beings; politics, on justice, etc. The upward direction is holism, seeking truth from systems and interactions, which is a broad road with boundless prospects. The downward direction is corpuscularianism. To explore truth by studying fireworks and particles is a narrow path —and there is no way out or up.

This is because the foundations of these disciplines reflect the nature of things. When the seeker looks down from the basics, he is looking downward from the essence of things and sees details and false identities whose endpoints are genes, quarks, neutrinos, etc. — where the differences of all things disappear, and a high degree of unity is achieved. But the distance from nature is also the greatest, and the farther it is from human purpose.

I'm not against reductionism and corpuscularianism, I just don't approve of using them as an arterial road for scientific research. I always thought that the path of holism, which is based on systems, life, and human beings, is the sunshiny road of science. And that's the path I'm following.

Two Entities of the Universe: Energy and Matter

There are two main points of view about the nature of the universe. One is matter ontology: the universe is composed of matter, matter is the only entity of the universe, and energy comes from matter. Another is energy ontology: the essence of the universe is energy, which is the ultimate source of all things.

The mainstream idea of Western culture is matter ontology, of which the atomism theory and particle theory are typical. Energy ontology is a new theory I have founded, that has not been put forward before. The theory closest to energy ontology is **Energetism,** a theory founded by Friedrich Wilhelm Ostwald, an early-20th century German physicochemist, which holds that energy is a more fundamental entity than matter and is the basis of all natural, social and mental phenomena. All these phenomena are manifestations of energy and its transformation and should be described and explained as processes of energy. He therefore advocated the exclusion of the concept of matter from science. He believed that matter and spirit were different forms of energy and could be transformed into each other. Therefore, the cosmology could be established only with the concept of energy, thus overcoming the traditional mind-matter dualism. But it is hard to support because after the discovery of Brownian motion, his position changed.

Quantum theory is another theory close to energy ontology, but its concept is vague, and its foundation is not solid. It is not yet a mature theory. Most importantly, no quantum scientist has explicitly linked quantum theory with energy theory yet, which is only my personal opinion. What they have been trying hard to do is to unify quantum theory with relativity, while relativity is not an energy theory either. The above description generally applies to **string theory**.

The basis of energy ontology is light field theory. It has the characteristics of unity and continuity, emphasizing energy– mass alternation. Light is the essence of the universe. The universe begins with light and ends with light. All flying particles are light. The photon is the unity of wave and particle and also the unity of energy and matter. The speed of light is the ultimate speed of the universe. The photon's motion and stationary state determines whether it is energy or matter. In an absolute sense, motion under the speed of light is neglected.

The Six Propositions of EBT (Energy-based Theory)

First, **the universe consists of energy & matter. They're transforming into each other all the time.** It is more realistic than the ancient Greek atomists believed when they claimed that "Nothing exists except atoms and empty space; everything else is opinion." (Democritus)

Second, **energy is the state of motion of light, free light, and all particles are photons in different forms.**

Third, **matter is trapped corpuscles** (photons, quarks, electrons, protons, atoms, molecules...), namely, trapped energy; While **energy is the free corpuscles**.

Fourth, **cosmic energy and material energy change with each other, excluding all intermediary.**

Fifth, **the basic form of particles is a field. It is continuous and unified.** The state of separated particles is the state when particles interact with matter. Particles are a continuum of energy & mass.

Sixth, **energy does not take up space.** The energy field always moves at the speed of light in the form of waves and interacts with everything. It is not itself conserved or occupying space. The carriers (atoms, molecules and objects) of material energy take up space.

The Principle of Division and Union

A basic principle of cosmology: material polymerization generates energy, and its decomposition consumes energy. This is abbreviated as the 'split–join principle'.

Particles polymerize into atoms, atoms polymerize into molecules, and molecules polymerize into everything. All of these polymerizations generate surplus energy, while their splitting consumes energy. This is a common phenomenon in the laboratory and even more common in reality.

In *New Guide*, Asimov give a typical example. "Hence, hydrogen gas almost always exists in the form of a pair of atoms—the hydrogen molecule. To separate the two atoms and free them as atomic hydrogen takes a good deal of energy. Irving Langmuir of the General Electric Company, who independently worked out a similar scheme involving electrons and chemical behavior, presented a practical demonstration of the strong tendency of the hydrogen atom to keep its electron shell filled. He made an atomic hydrogen torch by blowing hydrogen gas through an electric arc, which split the molecules' atoms apart; when the atoms recombined after passing the arc, they liberated the energy they had absorbed in splitting apart, and thus yielded temperatures up to 3400° C !" (Asimov, 1993).

Generally speaking, **decomposition is the destruction of existing products, but aggregation can create new products.** Fortunately, aggregation movement in the universe is far more common than decomposition movement. Aggregation is absolute and eternal; decomposition is relative and short- lived. Decomposition brings not infinite decomposition but new polymerization, because decomposition has its limit, namely, particles, while the eternal vortex motion makes the polymerization of matter & energy become infinite. This is the guarantee of the evolution of the universe.

The reason why energy is generated by material

polymerization has been discussed in the previous sections on vortex theory, and now we supplement it from another angle. The state of aggregation of matter is a field, not only particles will aggregate into tiny atom and molecule fields, but also large objects aggregate into fields, though the latter fields are relatively loose. They are all material fields, existing in the light field, and can somehow obtain EEEF from it, which constitutes another source of polymerization energy.

According to the foregoing description, I divide aggregation movement into three levels. The aggregation of particles is the combination of totality, their individuality no longer exist, and its force, the resonance force, is the strongest. The aggregation of atoms and molecules is a partial combination, usually there is only vibration without resonance, so it is not as powerful as the former combination. The energy generated by the aggregation of objects is the weakest.

Three quarks aggregate into a proton that belongs to the first-order binding. It does not need messenger particles (gluons) bound to it. They are twisted together, is unbreakable. The proton is no longer isolated three quarks combined but is the most basic and stable field in the universe, many scientists attempt to split it have ended in failure. Many scientists fail to see clearly a proton structure through a high-powered microscope for the same reason.

Protons, neutrons, and electrons polymerize into atoms, which belong to the second-order combination. Here only electrons combine into a unified electric field, while nuclei remain independent. The stability of this combination is far less than that of the former, but it is much stronger than the aggregation of objects in this respect. However, do not forget that vortex force is the greatest aggregation force, above all forces.

Material splitting usually takes place at high temperatures. The ionosphere of the earth is formed under strong sunlight radiation. Reducing water to hydrogen and oxygen also takes a lot of energy and is the manifestation of energy dissipation. This is because the decomposition activity is not simply the activity of releasing individual bondage but is also the activity of splitting the energy field.

The surplus energy produced by material polymerization comes from the particles themselves. Free particles are energy. After polymerization, they lose their individual freedom, and the force from this reaction partly appears as surplus energy, such as atomic vibration and molecular Brownian motion. Just like if we put a group of wild animals in a cage, although they lose their freedom, they still don't lack the power to shake the cage.

Using the split–join principle, the Big Bang theory and expansion theory are one-way decomposition theories — theories of continuous energy consumption and disappearance of the universe — whereas the theory of the formation and evolution of the universe belongs exclusively to the vortex theory.

SUMMARY

Polymerization is the natural aggregation of energy, the process of forming matter, or the process of imprisoning energy, without the need for additional energy. Since energy is speed, even in captivity, there will be slight movement, which will appear as particle vibration and excess energy. In addition, polymerization can often obtain additional energy from the light field.

Decomposition is the process by which energy is freed from captivity. It needs to break through the multiple shells that hold it. Energy inside matter that is not yet free can't do work. It needs extra energy to liberate it for it to work, just like an atomic bomb has to be detonated with orthodox explosives.

The Atomic Bomb Is Also a Fusion Weapon

Atomic energy is the first kind of cosmic energy discovered and developed by human beings. It's hidden in atoms and can be released in two ways: nuclear fission and nuclear fusion. The corresponding physical objects are the atomic bomb and the hydrogen bomb. The statement about the atomic bomb is not appropriate. It is a nuclear fusion weapon rather than a nuclear fission weapon in terms of destructive power.

According to the split–join principle, the energy produced by fusion is certain, whereas it is impossible for fission to produce extra energy; it can only consume energy, or the energy it consumes is greater than that of polymerization. The atomic bomb is essentially a cold weapon.

The reason for calling it a fusion weapon is as follows: the atomic explosion has both fission and fusion motions, and the fission motion is exaggerated, while the fusion motion is ignored. **It is true that the atomic bomb began as a fission movement, but it was followed by a much greater and longer fusion movement of at least two kinds.** One of these is that the ions released in large quantities fuse with atoms of air and dust to form new substances. Its fireball is the product of polymerization of uranium ions, and other ions, with oxygen in the air. The other is that the atoms that split apart are recombined. For example, uranium 235 is recombined into barium and technetium after fission. The fission movement is limited to the projectile body and its core, while the fusion movement extends to vast outer space and a large number of outer materials, including the atmosphere and the human body. **The energy produced by fusion far exceeds the energy consumed by fission, which is the source of the strong destructive power of the atomic bomb.** The same is true of incendiary bombs and detonators. If an atomic bomb is detonated in a vacuum, it will not cause shock waves or fires. No strong earthquakes or mushroom clouds follow. Its lethality would be no greater than that of an ordinary bomb, so If someone was there, he would be safe in concrete buildings thousands of meters from the centre of the explosion.

Cosmic Energy and Material Energy

Cosmic energy refers to the energy inherent in the universe. It is pure energy, the cosmic background field, which is invisible and uncontrollable to us. Cosmic energy transforms into matter and interacts with matter all the time, following the principle of equivalent conversion, meaning the total amount of energy in the universe always remains the same.

"Energy, in physics, the capacity for doing work. It may exist in potential, kinetic, thermal, electrical, chemical, nuclear, or other various forms...All forms of energy are associated with motion...**Energy can be neither created nor destroyed but only changed from one form to another. This principle is known as the conservation of energy or the first law of thermodynamics**...In the International System of Units (SI), energy is measured in joules. One joule is equal to the work done by a one-newton force acting over a one-metre distance." (Encyclopedia Britannica)

This is a popular definition. What is described above is actually material energy, which is the indispensable energy that we use every day. It is the basic belief of CFT to claim that **all matter & energy originate from cosmic energy.**

Material energy have material carriers, these types of energy can be measured and allocated, following the principles of energy invalidation and the law of the conservation of energy. Sunlight is a cosmic energy. When it is seen, sense, and utilized by us (which accounts for a small part of all sunlight), it becomes material energy that is often called solar energy. This phenomenon is often referred to as the transformation of energy forms, which is called energy- mass conversion in this book.

Material energy may be colloquially understood as the force that pushes and accelerates the motion of matter; it is not another material corpuscle, but the different ability of the same particles to move, with their motion appearing as speed.

The basic form of energy is a field. The cosmic energy is the light field, and the material energy is the particle field that is trapped in atoms, which are miniature fields. Then, once it is released, it is cosmic energy.

Generally speaking, light is the particle moving at light velocity — the speed of light — which is energy, and the particle is the photon moving at a lower speed, which is mass. Particle physics divides particles into physical

fermions and energetic bosons in a large way. They are two particle theories developed by two scientists that have now been linked. I agree with this approach and try to unify the two.

All particles originate from photons, so I define C as their boundary. C is the most prominent feature of the bosons, and fermions generally move below C. Thus, as long as the speed of the fermion reaches C, it realize the mass-energy conversion and become a boson. When the boson slows down, or stops moving, it transforms into a fermion.

This conversion is universal and occurs all the time, most typically in electrons and quarks. Physicists attribute them to fermions, which is expedient. Yes, when they lose their freedom and are trapped in atoms, they are fermions. But when they radiate from an object, or when they break away from an atom under voltage and become free particles, they are bosons. In addition, **the photon is recognized as a boson, but when it loses its freedom and becomes the building block of matter, it transforms into a fermion.**

Energy particles, mainly photons, electrons and neutrinos, are self-propelled and can fly at the speed of light; mass particles, including protons, neutrons and all composite particles, are driven by external forces, and their speed of motion cannot reach the speed of light, even with the help of accelerators

From this view, I get a comprehensive definition of energy and matter: **Energy is the motion state, free state,** and field state of the photon, while matter is the static state, confinement state, and agglomeration state of the photon.

Energy is the force that propels and accelerates the motion of things, matter is the entity that transforms and absorbs energy. The equivalent transformation of energy and matter is the most basic variable motion of the universe.

The universe is made up of energy and matter. Matter is the first object of scientific research, having produced heliocentric theory, atomism, Newton's laws, physics, relativity and other major achievements. Scientists have also done a lot of research on energy, producing dynamics, thermodynamics, electromagnetic theory, the four natural forces theory, quantum theory and other achievements. As these theories are guided by reductionism, they are scattered and one-sided, and lack of an internal coherent connection.

We say that the universe is one, like a well-designed system, and that everything visible and invisible, up to man, is an interrelated whole. **It is the sacred task of philosophy and science to find this connection and to establish a unified discipline, including the science of the unification of energy as well as matter, and the cosmology of the unification of heaven, Earth and man.** Through this discussion, I believe that we have taken a great step forward along this road.

Cosmic Field and Material Fields

The basic form of energy is field. Corresponding to the two energies are two fields: the cosmic field and the material field. The cosmic field is the field of light. Although it is derived from matter, but compared with its space, matter is insignificant and can be ignored, so we can define it as a pure energy field, an infinitely expanding space filled with light ether. Any movement can get EEEF from it, more or less, conspicuous or concealed.

Light is the element and medium of light field, or light ether, and its eternal motion does not depend on other medium. The field of light is vast, enveloping all fields of matter, and it is outwardly divergent and open.

The material field is composed of continuous matter, from the miniature proton field to the super galaxy field is a series of matter motion. The common feature of their motion is inward convergence, closed, need medium. The fields of protons, atoms, molecules, and objects have only internal motion and no obvious external motion. The fields composed of air, water, celestial bodies and so on have obvious external motion, and the velocity of the upper level is faster than the lower level. Starting from the ground, we can see that the ocean doesn't move as fast as the air, which is slower than the Earth orbit velocity, that, slower than the sun's orbit velocity, that, slower than the Milky Way orbit velocity, that, slower than the local galaxy orbit velocity... On this basis, astronomers establish the first, second and third cosmic velocities.

The main function of field is to transfer energy in the form of waves. Four common fields are discussed in detail below.

MAGNETIC & ELECTRIC FIELDS

A magnetic field is formed when cations vibrate together. It comes in three main forms: natural magnets, geomagnetism and electromagnets.

A natural magnet is a kind of metal in which the plasma failed to reduce to atoms as the earth cooled, and its anions & cations are separated. These metals are mixed up with neutral atoms, so they are not very magnetic.

The geomagnetism is caused by the Earth's spin. This motion causes the air above the Earth's poles to spiral in one direction, creating two low-pressure centers that absorb air and corpuscles in opposite directions, **turning atoms into electrolytes. All spinning celestial bodies thus generate a magnetic field whose strength is proportional to its spin velocity.** Mercury, Venus, and the Moon spin so slowly that their geomagnetism is so weak that astronomers are almost impossible to measure it. Jovian stars spin much faster than the sun, and even though they have much less mass comparing to the sun, their magnetic fields are much stronger than the sun's. The black hole has the highest spin velocity and the strongest magnetic field of any celestial body. **The geomagnetism suggests that the magnetic force does not come from the mysterious 'domain', but the vibration force of electrolytes, most cations.**

Electromagnet is artificial magnetic field. **Electricity is the unity of opposites between the negative electron field & the positive ion field, an electromagnetic field in itself.** By adjusting devices, controlling its A-I cycles and making the machine resonate with it, work can be done.

WATER FIELD

It usually transmits energy in the form of water waves on the surface, proportional to the mass of the energy-carrying object. The previous example of throwing stones into water shows that if we throw harder or throw a bigger stone into the water, it will arouse a higher and larger wave, that is, waves of shorter wavelength and higher frequency, and vice versa, a longer waves and lower frequency. In both cases, the central wave is highest and strongest, and the waves get lower and weaker as they go outward. In this case, the energy carrier—the moving stone—stops moving as soon as it sinks but passes its energy around in the form of accelerating the motion of the water molecules, which simply fluctuate up and down without moving off their positions.

If the energy carrier is in motion, i.e., the source of the wave is in motion, the water field follows it. For example, a ship on the ocean, its bow with strong impact in front of the sea water, arouse tall waves, its wavelength is short, and frequency is high; behind, the wave is relatively gentle, lower frequency longer wave, which can be called the fluid Doppler effect. The capitals of most ships are designed to be conical in order to reduce bow resistance.

AIR FIELD

The most prominent is the field motion induced by sound, which is related to the direction of motion of the sound source object. Radio and television speakers are always designed to face the listener directly because the sound waves on this side are higher in frequency and shorter in length than those on the side and back, and because they push more air molecules around faster, the listener can listen better. If the machine is moved towards the listener, the sound is enhanced, which scientists call **the (air) Doppler effect.** When a train or a car is in motion, the Doppler effect is stronger when the whistle and horn go on. Their bodies drag some of the air in the same direction, but their sound only compresses the air, not pushes it to move much. When the vehicles pass, they will return to their original position.

The propagation of sound in the air is positively related to the temperature and density of the air, which is a kind of speed that varies with the environment. Sound travels at 340 meters per second in air at 15°C. The airplane is a fast-moving sound source, creating the Doppler effect all the flying time. When the speed of the aircraft is in sync with the speed of its own sound waves specified in the environment (slightly lower than 340m/s, due to it happening at higher altitude), the compression waves cannot speed away from the plane. They are piled up in front of the wings and body of the aircraft, forming a veritable wall called **a shock wave, i.e., a sound barrier.** Modern aircrafts are designed to be supersonic, able to travel through the sound barrier without being harmed by shock waves.

LIGHT FIELD

The light field is analogous to the air field in the above example. It is uneven. The closer it is to the light source, the greater its density. The sunlight starts to radiate outward from the place with the highest density of the light field, and its speed is lower than the speed of light. When it flies out, the density of light field drops rapidly, and the density of the sunlight drops correspondingly, but its speed rises rapidly. When it reaches the speed of light, its frequency exceeds the frequency of light field, which mixed with a lot of exotic starlight, and light field becomes its resistance, that's similar to sound barrier and forms **light barrier, namely corona. The corona is the sun's first outer hot shell. Its EEEF is very powerful, thus it become the sources of solar wind and interstellar dusts. This is a feature common to all stars.**

The above discussion can be extended to an even more kinky question: the origin of EEEF. We need to recall an idea about light made by Pietro Angelo Secchi, the 19th century Italian astronomer and physicist famous for his work on solar phenomena, stellar spectroscopy, and spectral classification.

"The fact is, the sun's rays have different effects according to the length of the vibration: those with the greatest wavelength communicate heat, those with medium wavelength, light, and those with the shortest wavelength, chemical action." (Engels, 1883-*Dialectics of Nature*)

As today's high school students know, sunlight covers the full spectrum of light. Among them, red light has the longest wavelength and the lowest frequency. It cannot enter the interior of most objects and only interacts with the particles on the surface of them. Violet light, which has shorter wavelength and higher frequency, can easily penetrate the surface of an object, enter its interior, and interact with its internal particles. White light is somewhere in between, omitted here.

Since the establishment of particle collider, particle collision experiments have been carried out frequently and many achievements have been made. No experiments have been done on the collision of photons of different wavelengths and frequencies (Tip. The light field does not fit the collider), and I believe similar results will be achieved from it. When red and violet light impinge in the light field, the following event is surely to happen. **The violet light will penetrate into the red light. Part of the red lights will block the violet light from flying, forming light barriers such as the sun's corona and 'fire wall'. The other part of the red lights embrace the violet light, forming ions and atomic stuffs, such as the earth's ionosphere, ozone layer and atmospheric molecules.**

Due to the light field is extremely rich in resources, its output is far greater than its input, hence the EEEF. It's also possible that inputs and outputs are equal, and we just don't see the inputs, they are energy.

It should be noted that a star radiation is uniformly outward, and there is fewer impinging itself. This event mainly occurs among stars. For example, when Andromeda starlight reaches the solar system, its wavelength must be much longer than that of sunlight (Reference. 'Light Waves Get Longer as They Travel and All Starlight is Redshifting'), and its red light will become a barrier to the sun's violet light and form material, which is the main source of interstellar dust and nebula.

A. Secchi had asked: "When the sun and the whole system are extinct, are there forces in nature which can reconvert the dead system into its original state of glowing nebula and reawaken it to new life? We do not know."(Ibid)

I think the above is the best answer so far.

GUT of Energy Achieved by Particle Vibration

Nikola Tesla once said, "If you want to find the secrets of the universe, think in terms of energy, frequency, and vibration." As we look at his life and work, we'll see that he did just that.

Justenergy.com

"Scientists are discovering now that everything in the Universe vibrates at one frequency or another. That includes human beings...Ancient wisdom, and many indigenous people today, bear witness to the fact that even inanimate objects have vibrational energy. Even modern scientific findings, such as superstring theory, support the idea that **vibrations are at the core of all existence.** The vibrations are there, but in modern culture we are not taught to hone the senses that can detect this type of energy." (Jocelyn Mercado, Pachamama Alliance, 'Becoming Aware Universe', wakingtimes.com)

"The invisible electron and the light of the sun can be described and harnessed with Maxwell's electromagnetic laws. Electric currents, magnetic effects, radio waves, infrared waves, light waves, ultra-violet waves, X-rays, and gamma rays, sinusoidal waves with frequencies as low as 60 per second and as high as 1 followed by 24 zeros are manifestations of one underlying mathematico-physical scheme. This theory, which is at once so profound and so comprehensive that it beggars the imagination, has revealed a plan and an order in nature which speaks more eloquently to man than nature herself." (Kline, 1953, p.316)

Cosmic energy, free light, is infinite, and only a tiny part is imprisoned by the material shell and becomes material energy. However, free light is beyond our control and utilization. Reasons: 1. Invisible. It is the light itself, we can sense it but can't see it. All we can sense & see is the interaction between light and things. 2. Low density. It is the background temperature, 2.7K, measured by scientists. 3. High speed. It always flies outward from its source at the ultimate speed.

Material energy is also invisible, but its outer shell is visible. They are often grouped together to form a material field in various forms, such as solar energy, wind energy, water energy, electrical energy, magnetic energy, chemical energy, and so on that we see and use every day, measured in joules, watts, calories, etc.

In the last century, scientists' ambition was to unify the four natural forces (energy), which was called GUT (Grand Unified Theory) and TOE (Theory of Everything). As a result, the unity of electromagnetic force and weak force was realized mathematically, and the GUT failed, TOE was even more unlikely. The fundamental reason is that they lack a unified basis.

CFT has this foundation, that is, light and vibration. It can achieve unity. We probe this issue now.

Particle physics tells us, all things are made up of particles, and a common characteristic of all particles is vibration. The object made up of them is always in vibration, sometimes in motion, except at absolute zero, a temperature that will never occur.

A basic view of CFT is that all particles originate from photons, which are free energy or confined energy. **Because shells are everywhere, the flight of particles is restricted, and part of their energy is generally converted into vibration, whose strength is determined by its degree of freedom.** While the stars and planets orbit, they spin and vibrate widely; Molecules and cells vibrate slightly while moving randomly; Atoms are always vibrating slightly in situ and violently as they lose electrons to become ions; Photons vibrate at the highest frequency and shortest wave while flying at high speed. Everything is vibrating and moving, and even the stationary things are always vibrating within.

Vibration is a small and repeated rhythm of all

objects, which is a process and presents a wave shape. Each wave can be described by two indices, wavelength (λ) and frequency (f), and has a relatively low starting point, low end point called troughs, and peak crest. The general rule is that large & heavy objects vibrate with longer wavelengths and lower frequencies, then the wavelengths are shortened and the frequencies are increased in order. The periodic table of elements has this feature.

"Then, in 1914, the young English physicist Henry Gwyn-Jeffreys Moseley determined the wavelengths of the characteristic X-rays produced by various metals and made the important discovery that **the wavelength decreased in a very regular manner as one went up the periodic table.**" (Asimov, 1993)

All things can vibrate & move, which is the embodiment of their vitality. Inanimate things neither vibrate nor move. **I divide particles into three categories according to their freedom: Energy particles, semi-mass-semi-energy particles and mass particles. Photon, ion and atom are the representatives of the three.** They form light field and particle field with different frequencies of vibration.

Any motion consumes energy. In CFT, all particle own its innate inexhaustible energy. It is an eternal energy, because its essence is light, and light is eternal. **All particles are eternal, and so are their vibrations. They change oft, but never die. They are the source of all our daily energy. The core subject of applied science can be summed up as the study and utilization of material vibration. Particle energy is invisible, but particle vibration is relatively visible, or sensible and measurable, and scientists did it, with extreme precision.** In light, for example, scientists have determined precisely not only the wavelength and frequency of white light, but also those of various color lights. In fact, scientists have used this method to achieve the grand unity of energy, including material energy and cosmic energy, namely, the existing electromagnetic spectrum table— I think the correct name for it is **light spectrum table.**

Matter is trapped particles, vibration & motion are common attributes of all things, such as microwave, radio wave, ultrasonic wave, sound wave, infrasonic wave, water wave, seismic wave..., and in this sense material vibrations could also be added to the red end of the table. And light spectrum table can be renamed **particle vibration spectrum,** actually TOE that scientists dream of.

According to this new TOE, CFT declares: **vibration is the uniform observable form of energy in the universe.**

The vibration energy of a single particle is very weak and can be ignored. Only the vibration of the particle field can generate considerable energy. The energy of a particle field is the collection of all particle vibrations, which may be destructive or constructive force, and it can be divided into two types. **When their vibrations are random, most of the energy is canceled out by each other, and the effect is small. When they vibrate uniformly & orderly, the effect is much stronger, even scary, which is called resonance.**

Considering people as particles, the vibration force of individuals is too small to destroy a bridge, but when their vibrations resonate, they can do it. There is such a typical example in history.

"In 1831 a suspension bridge in Greater Manchester, England, collapsed while British troops marched over it. The bridge had a few design flaws, but the marching of troops was discovered to be the reason for the collapse. When the 74 soldiers marched back, they crossed the suspension bridge, four abreast. As they marched, they noticed the bridge had begun to vibrate in time with their steps. They were not aware that they were inducing mechanical resonance within the bridge, which would lead to a catastrophic structural failure. After this incident, the British Army put out a new order that marching troops should 'break step' when crossing a bridge.

Mechanical resonance. All objects, even ones that

appear solid and stationary, like buildings, walls, and bridges vibrate at various frequencies. **If a force is applied to the object that matches its natural frequency, it will amplify its vibration. This phenomenon is called mechanical resonance and has caused a number of high-profile disasters."** ('The British Army stopped walking in-step on bridges because of an incident in 1831' by Jesse, Guest Author of warhistoryonline.com, Sep 28, 2021)

"Resonance, in particle physics, an extremely short-lived phenomenon associated with subatomic particles called hadrons that decay via the strong nuclear force. This force is so powerful that it allows resonances to exist only for the amount of time it takes light to cross each such 'object.' A resonance occurs when the net energy of the colliding subatomic particles is just enough to produce its rest mass, which the strong force then causes to disintegrate within 10^{-23} second." (Encyclopedia Britannica)

"All objects, not just fundamental strings, have resonant patterns associated with them. Pluck the string of a violin and you hear mainly one tone. This is the string's fundamental resonant pattern, or frequency. And the instrument's resonance doesn't stop there. The body of the violin has resonant frequencies, which work to amplify the sound created by the vibrating string. **There's resonance in objects that aren't musical, too. Your desk has resonant frequencies, and so does a flagpole, and so does the Earth."** (The Elegant Universe homepage)

From corpuscularianism, scientists have discovered more than 100 short-lived resonant particles that seem to have little to do with our daily life. In the holistic view, the continuous and even eternal vibration & its resonance motion is everywhere. Stars are the eternal powerful ion vibration balls, and electricity is the intermittent powerful ion resonant force, they brings us steady streams of light & heat.

Radiation Is a Universal Characteristic of Objects

That every object radiates a power (species) by which it acts upon nearby objects suited to receive those species.
    Roger Bacon (https://books.google.com)

"Radiation may bring to mind the superheroes and monsters of comic books and movies, **but radiation is very real and all around us! In fact, you are currently being bombarded by radiation.** It might be coming from the sun, various electronic devices you own, or even the food in your kitchen. If you have ever eaten a banana, you have eaten a radioactive material. The good news is that the vast majority of radiation you are exposed to is relatively harmless...So the difference between ionizing radiation and non-ionizing radiation is that **ionizing radiation has enough energy to strip electrons off of atoms, and non-ionizing radiation does not have enough energy to strip electrons off of atoms.** One of the easiest ways to visualize the difference between these two is to look at the frequency spectrum for light. As the frequency goes up, so does energy, so we can see the energy cut off for light where it goes from non-ionizing to ionizing radiation is within the ultraviolet light spectrum." ('What is Radiation? Definition, Causes & Effects'. By Damien Howard & Christianlly Cena, https://study.com/academy/lesson/)

Everything comes from the environment, they cannot be separated from the environment, and have to adapt to it. But the adaptation is not entirely passive, there is a general and active act, namely radiation. It is a mechanism by which all things preserve themselves by changing the environment to some extent.

Everything is radiating. Radiation from atoms is called decay; from gases, diffusion; from liquids, volatilization; from organisms, heating; from stars, luminescence; from black holes, radiation is mistaken for neutron star's pulses. Finally, all free particles themselves are radiators.

This characteristic is first related to the property of the motion of objects. All things are moving at all times, including whole motion, parts motion, autonomous motion, motion pushed by external force, aggregated motion, and discrete motion. **Radiation is an autonomous and discrete motion of parts (particles, atoms, molecules) that occurs at all time. Secondly, it is related to the cosmic background temperature, the absolute zero, with all things being hotter than it is. According to the second law of thermodynamics, all things must always be radiated.** Radiation is a common characteristic of objects. The only differences in radiation between objects are in strength and form.

Here, the second law of thermodynamics can be described as: **stronger radiation always suppress weaker radiation. CFT applied this law in another direction to establish the shell theory.**

Compared with radiation, gravity is so weak that the force of radiation alone can effectively prevent celestial bodies from clumping together under gravity.

Radiation is outward, from the core to the surface, from the surface to space. **This is repulsion, another innate force of matter, which exists from particles to stars.** Its intensity is proportional to the core pressure, can be explained by using Newton's third law. Radiation is anti-pressure, end product of tension. But the pressure causes the object to contract, therefore the radiation intensity is also proportional to the object's contraction force.

Stars are the most obvious example. If a star is expanding, it is a sign of pressure drop and radiation attenuation. On the other hand, if it is contracting, it is a sign of increased pressure and radiation. The three factors can be causally related to each other. For example, when we observe that the radiation of a star is gradually increasing, we can conclude that it is contracting, and the internal pressure is increasing. Jupiter is a typical example.

The radiation of all things, especially celestial bodies, is a kind of antigravity which is not constrained by action at a distance. Radiating celestial bodies increase the distance between each other, while the body itself receives external radiation that strengthens its own gravity. It's like two people shooting water at each other. The offset of the two water streams weakens the impact of each other, stretching the distance between them, and adding weight by absorbing the jet of water from the other.

"Non-ionizing radiation is limited to the lower energy range electromagnetic radiation, which is more commonly known as light." (Ibid.) Sunlight is composed of photons and ions, i.e non-ionizing radiation & ionizing radiation (the solar wind), and the latter that mainly interacts with matter. It creates the corona, which powerfully preserves its energy; It creates the earth ionosphere and thermosphere, effectively preventing the loss of atmospheric and ground objects. It helps plants grow, make oxygen and chlorophyll, and human beings cannot live without the sun.

For the solar system, scientists in recent years also found a high temperature 'fire wall' in the space 120AU away from the sun, that is, the fire shell enveloping the entire solar system, effectively blocking the escape of radiation material, making most of them trapped in the solar system.

These facts show that **radiation is the only mass and energy flow that can be transmitted indefinitely without being limited by the action of distance. Since vibration is the property of all mass and energy, radiation is also a source of new stuffs.** The two most obvious ones on Earth are the ozone layer and the auroras, which are products of the radiation and vibration of sunlight.

Despite all the interception of celestial radiation, the total amount of radiation in the universe is still staggering. If we could see it, it would be a circular radiator 100 billion light-years across, expanding infinitely to outer space at a speed of 600,000 kilometers per second.

SUMMARY

1. The pure form of cosmic energy is the light field; Its common material forms are vibration and radiation, called material energy. It is infinite and eternal, only changing forms and never decaying.
2. Vibrations & radiation are properties of everything from particles to stars, without exception. That's their forces.
3. Vibration & radiation are essential & inherent to matter, not external forces. They are spontaneous and will always be generated even if there is no interaction.
4. Vibrations & radiation always transmit energy in the form of waves. The intensity of a wave is determined by its wavelength (λ) and frequency (f). We know from the X-ray series of the periodic table that the elements are arranged in the order of long wavelengths-low frequencies to short wavelengths-high frequencies.
5. Vibrations are divided into three classes according to their characteristics and intensity: energetic particle (photon) vibration, semi-mass-semi-energy particle (ion) vibration, and mass particle (atom, molecule...) vibration.

    Comparatively, the photon vibrates with the shortest wavelength and the highest frequency. Atoms are at the opposite, and ion is in the middle. Vortex is the mixed motion &vibration of three kinds of particles, mainly ions, which move & vibrate far more violently than that of mass particles & bodies, causing stars to glow, spin and orbit.
6. Vibration & radiation usually form material fields whose strength is determined by three indices,λ, f, d (density). In the above three fields, the micro fields inside atoms and molecules are very solid, but they are generally jumbled and scattered, and

therefore weak. The light field, though composed of the shortest wave and highest frequency of photons, is weak due to its high diffusion and overall low density. The strongest of the three is the ion field, which is represented by the dazzling stars that fill the sky.
7. Resonance is a strong field motion, requiring higher frequency and higher density, which is the universal characteristic of ion field. Stars, incandescent lamps and lasers emit resonance light, and the shock wave of nuclear bomb and EEEF of light field are essentially resonance effects.
8. Vortex is the mixed motion & vibration of various particles, mainly ions, which move & vibrate far more violently than that of mass particles & bodies, causing stars to glow, spin and orbit.

The Mechanism of the Self-Preservation of Everything: Shells

Radiation is, on the one hand, an act of aggression, a conquering force. On the other hand, it is a kind of separation force, letting the object lose energy and matter, and start to decay.

One approach to combat loss and death is to aggregate movement to extract energy and matter from the environment and extend life. Another approach is to build multiple shells to slow down the radiation.

Shells are divided into hot shell and cold shell. Ordinary objects, including atoms, molecules, cells, animals and plants, are mostly wrapped in a cold shell. It is also composed of particles, that act as the surface layer of an object that has no other outer shell. Compared with the interior of an object, its density is higher, the distance between particles is shorter, and the moving speed is limited, so its temperature is lower.

All celestial bodies are wrapped in shells. **All stars are wrapped in hot shells,** determined by the velocity of their own ions. This feature is crucial for stars. Stars are strong radiators. If they have a cold shell instead of a hot one, their energy and matter will soon run out and disintegrate.

A star is a strong radiator, and its shell must be a more powerful radiator to protect it. This is based on the second law of thermodynamics, which I'll rewrite for the sake of argument: Strong radiation suppresses weak radiation, and short-wave radiation invades long-wave radiation.

We know from the geological records of the Earth's meteorology that the temperature of the Sun is stable, and there is no trend of continuous warming or continuous cooling, which can be attributed to its hot shells.

CFVT asserts that the core of the Sun is the lowest

temperature area, and outside it is the multilayer hotter shells, with the highest temperature in the photosphere, which is the strongest hot shell of the body. However, it is often pierced by sunspots, causing slight losses.

The corona is a generous thermal shell outside its body. Its million-degree temperature strongly blocks the radiation that has just broken through the photosphere and chromosphere, and the sun can survive for five billion years without changing its true colors — its inner and outer shells are indispensable. Other main-sequence stars are stable in the same way.

The Earth also has its shells. The earth's crust is on the ground, and the ozone layer and ionosphere are in the air. All of these play a protective role—They also create the greenhouse effect. Shells are automatically generated and cannot be abandoned. All things grow in layers of shells; they grow with the growth of objects. Once the shells are destroyed, the objects will die.

Ignoring the shell and lacking the knowledge of its functions is an important factor for astronomers to abandon vortex theory and stick to the theory of gravitation, in which no shell is required, and no shell is allowed. It is also the reason why Friedrich Wilhelm Ostwald and some quantum scientists deny matter.

The Main Force of Antigravity Is Radiation

In *A Brief History of Time*, Stephen Hawking wrote:

"Einstein introduced a new 'antigravity' force, which, unlike other forces, did not come from any particular source but was built into the very fabric of space-time. He claimed that **space-time had an inbuilt tendency to expand, and this could be made to balance exactly the attraction of all the matter in the universe, so that a static universe would result."** (p.47)

This theory may be a basis for cosmological expansion theory and Einstein's construction of cosmological constant, but it cannot be established in physics. Only matter and physical space show the phenomenon of expansion or contraction, which does not exist in pure space-time. Like the concept of vacuums, the concept of pure space-time is virtual and does not exist in reality. So, if material is completely excluded, it's the real vacuum, and such thing as expansion space, flat space or curved space cannot exist.

But antigravities are ubiquitous, one of them is radiation. According to the second law of thermodynamics, the temperature of all objects is higher than the background temperature of the universe, making radiation the most common phenomenon and the main force of antigravity.

"Radiation can be described as energy or particles from a source that travel through space or other mediums. Light, heat, and the microwaves and radio waves used for wireless communications are all forms of radiation.

Radiation includes particles and electromagnetic waves that are emitted by some materials and carry energy. The kind of radiation discussed below is called ionising radiation because it can produce charged particles (or ions) in matter. X-rays, gamma-rays, alpha particles, beta particles and neutrons

are all examples of ionising radiation."
(https://www.ansto.gov.au/education/nuclear-facts/what-is-radiation)

The radiation of all things lasts for a lifetime. The absorption is intermittent, and most of the radiation they absorbed comes from other things. External factors will affect radiation, such as the hot corona would severely suppress the sun's radiation, but can't completely block it.

Isolated radiation is the loss of material and energy from objects, while bidirectional and multidirectional radiation may be loss and compensation, which is related to distance. If the distance between the two sides is long, the loss is large and the compensation is small. If the distance is close, the situation is opposite. When the two sides are in contact, molecules and atoms bump into each other, which will produce interaction and combine into new molecules and atoms. This is the most significant gravitational compensation. However, compared with radiation, compensation is limited after all. Most radiators are dominated by antigravity, and gravity is inferior or even non-existent. Generally speaking, vortexes and black holes are gravitationally dominant bodies, and stars are antigravitational dominant bodies — otherwise known as the contraction body and radiator respectively.

Gravitation is suspicious. It has insurmountable obstacles; that is, action at a distance. According to the tug-of-war model, gravity is actually antigravity. Antigravity is intuitive and definite; where there is gravity, there is an equal antigravity, with no room for doubt and no problem of action at a distance. I find it hard to understand how Newtonians generally ignore radiation and other antigravity forces. **Antigravity is a necessary condition to prevent everything from forming a big block and keeping their independence.** The stability and harmony of the universe is not so much the function of gravity as it is the function of antigravity. This is like human society maintaining its stability. Human beings have always been fighting rather than loving each other. Human beings exert force and output energy in everything they do, even when one is pulling an object toward oneself. This can also be understood as antigravity and radiation too.

According to Newton's third law, each force has an equal and opposite force on a line. This is the basis of the concept of antigravity. I guess that Newton considered establishing the concept, simply because he could not determine the nature of gravity, so antigravity was out of the question. Einstein replaced gravity with curved space, and antigravity should be straight motion, inertial motion, the effort to restore flat space, and endeavour to recede from countless pits created by heavy objects. Wow!

Magnetism and Gravity Disappear at High Temperatures

It is a fact of science that magnetism disappears at high temperatures and no longer needs to be proved experimentally. It is also common sense in science that the molecules & atoms of an object in ordered state to form a magnetic field. High temperature means that the motion of molecules & atoms is accelerated, and the result is to break the order, the motion become disorder and chaos, and the magnetic force disappears.

A closely related question then arises: **Could gravity also disappear at high temperatures? According to Newton's law, the answer is nope. If yes, then stars would have only antigravity and no gravity at all. Only planets and moons have this ability.**

Because gravity is weak, it is difficult for us to get different answers through experiments, so we can only continue to use logical reasoning.

The reason is simple and clear: heating is the acceleration of corpuscular motion, which causes objects to expand, the distance between molecules & atoms increases, and the number of escaping molecules & atoms (radiation) increases. All these are facets of antigravity. To a certain extent, these factors will overwhelm gravity. If the temperature continues to rise, the molecular & atomic object is transformed into a complete radiator, what gravity can there be?

Therefore, I agree with astronomers' observation of magnetic field, but not their observation of gravity.

They point out, the mass of the Sun is 330,000 times the mass of the Earth, but its magnetic field intensity is only twice as high as that of the Earth — only 1-2 gauss in its polar field. (NASA. Sun Fact sheet) This result can be attributed to high temperature. The strong magnetic field of sunspots is caused by vortex motion, which is discussed elsewhere in this book. As for gravity, it is directly proportional to the mass of an object. But scientists have not determined the surface gravity ratio of sun/earth at 330,000, but only 28. (NASA. Planetary Fact sheet) I even deny this ratio, **because I have renamed gravity as the aggregation force of particles, which can only be produced by mutual contact & resonance.** Besides, the sun is a powerful radiator, and even if we ignore the action at a distance and admit its gravity, the gravity is so weak that will be eliminated in the sun's strong radiation, or the high temperature.

New Mode of Gravity: Tug-of-War

Newton's great generalization, which he called the 'third law of motion', was that 'Action and reaction are always equal to each other'; and that law has been one of the most pregnant of all truths about the mystery of force; — one of the brightest windows through which modern eyes have looked into the world of Nature.

Phillips Brooks (GIGA Quotes)

Force is interaction. This is Newton's definition of force. It shows that force does not exist in isolated objects but is the result of the interaction of more than two objects. An isolated object has only energy, no force.

In the sky, all scattered celestial bodies experience weightlessness, as there is no force, and only show weight and force when they interact with each other. If we ignore the reaction, all celestial bodies attract each other and that the phenomenon of clusters will appear, which is called Bentley's paradox. I'll give it an romantic name: hug mode or lover's impulse. Facts have proven that this phenomenon does not exist. Instead, there is a phenomenon of mutual confrontation, similar to tug-of-war in sport. The fable of *The Swan, Prawn, and Barracuda Pulling a Cart* is also suitable here, which illustrates the confrontation between gravity and antigravity in space.

Principle: **Every object is a gravitational center, and it must pull any peripheral objects toward itself. The direction of force that an object and any peripheral objects exert is opposite each other, like monopole magnets** (Newton's load- stone is bipolar magnet), which will effectively prevent the phenomenon of agglomeration. In addition, everything is a radiator and will resist gravity.

In the case of a tug-of-war on Earth, both sides are down-to-Earth, and the strong team can pull the weak team over the centre line and win. The tug-of-war in the sky will not produce such a result due to all parties are weightless, there is no solid fulcrum, and there is no place to exert their strength. Both sides will be deadlocked on a balance point for a long time. Their combined strength produces not a state of mutual attraction or overall rigidity but a state of mutual rotation. The speed of rotation is inversely proportional to the mass, which is the inevitable result of the imbalance of counterforce. This is shown by the slow rotation speeds of heavy objects and the fast rotation of light objects around heavy objects, such as the moon rotating around Earth. This kind of scene is often seen in the skating performance of a man and woman too.

In fact, the tug-of-war mode is the application of Newton's third law of motion. Newton had given many examples to illustrate it, and one of them is quoted below: A horse vs a stone.

"If a horse draws a stone tied to a rope, the horse (if I may so say) will be equally drawn back towards the stone: for the distended rope, by the same endeavour to relax or unbend itself, will draw the horse as much towards the stone, as it does the stone towards the horse, and will obstruct the progress of the one as much as it advances that of the other." (*The Principia*,14)

Why didn't he apply this law to celestial relations? Obviously, he had been puzzled by the action of distance, there is no 'rope' (hand in hand in the pairs skating) between celestial bodies.

NOTE. Recently, for the second edition of this book, I read *The Principia* again and found that Newton didn't ignore his third law, just he attached it to *attraction* rather than gravity. To Newton, attractive force (centripetal forces, impulse) is not originated by gravity, that is, gravity is not the cause of attraction, it is just the common center of the mutual attraction of celestial bodies. It is also the common center of both sides of the tug-of-war, but the directions of exertion is opposite.

"SECTION 11. OF THE MOTIONS OF BODIES TENDING TO EACH OTHER WITH CENTRIPETAL FORCES

I have hitherto been treating of the attractions of bodies towards an immovable centre; though very probably there is no such thing existent in nature. **For attractions are made towards bodies, and the actions of the bodies attracted and attracting are always reciprocal and equal, by Law III;** so that if there are two bodies, neither the attracted nor the attracting body is truly at rest, but both (by Cor. 4, of the Laws of Motion), being as it were mutually attracted, revolve about a common centre of gravity. And if there be more bodies, which are either attracted by one single one which is attracted by them again, or which all of them, attract each other mutually, these bodies will be so moved among themselves, as that their common centre of gravity will either be at rest, or move uniformly forward in a right line. **I shall therefore at present go on to treat of the motion of bodies mutually attracting each other; considering the centripetal forces as attractions; though perhaps in a physical strictness they may more truly be called impulses.** But these propositions are to be considered as purely mathematical; and therefore, laying aside all physical considerations, I make use of a familiar way of speaking, to make myself the more easily understood by a mathematical reader." (Ibid. 160)

A comparison of the two modes of gravity:

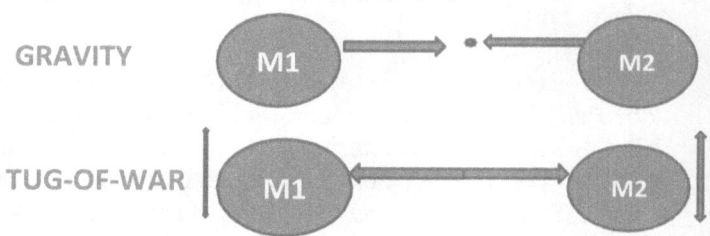

The two vertical lines of the tug-of-war mode are system boundary (shell). In most cases, the boundary is a spherical.

A Tug-of-War between the Earth and the Moon

In terms of the closest and most familiar relationship between Earth (M1) and the Moon (M2), they have a confrontational relationship, applicable to the tug-of-war model.

There is a balance point in their tug-of-war at which the two forces cancel each other out and enter a longer or shorter stalemate. If the two sides are even, this point is right in the middle of them. The Earth–Moon confrontation belongs to the strong and weak confrontation. The balance point will not be in the middle. We can only look for it on the side of the Moon. According to the ratio showing that the gravity of the Moon is one-sixth the gravity of the Earth, we can divide the distance between the two into six segments. Taking the Moon as the starting point, the equilibrium point between the two will fall on the intersection of the first and second segments. That is, the Moon occupies one segment, and Earth occupies five segments. At the equilibrium point, the two forces of gravity cancel each other out to zero, which astronomers call Lagrangian points. Using the previous ratio, the result is very close to the value of L1 provided by mathematicians.

All objects in space are weightless, which shows that there is no distance action, and there is no attraction between celestial bodies, only mutual repulsion force and random drift motion caused by radiation. Instead of falling apart, they run orderly and produce everything, which owe to the action of vortex motion and shell.

What is the earth-moon system's shell? I think it is the earth orbit. Don't think of it as a linear circle, it is a spiral tubular channel with a radius of 449.000 kilometers, in which the earth and the moon run. Here, I regard the gravitational equilibrium point L2 recognized by astronomers as the margin of the Earth-Moon orbital shell, which can follow the asteroid and artificial spacecrafts on it.

At a distance of 1,500,000 kilometers from the earth, mathematicians calculated L2 of the sun-earth system, which has the same function as the earth-moon shell. Astronomers applied this method to Jupiter, and found that there are millions of trojan asteroids at its Lagrangian points, viz., its shell.

Apparently, all orbiting celestial bodies have Lagrangian points, that is, they all have orbital shells. Therefore, the shell structure of the solar system is very complex and orderly, they are the products of vortex motion. Without shell structure, the stability and order of celestial system is unimaginable.

The balance point I'm talking about here is Newton's common center of gravity in *The Principia*, but they fall on opposite ends of the line segment. My balance point is on the side of light objects, Newton's common center is on the side of heavy objects. For example, the common center of gravity between planets and the sun is not on the other side of the planet, but near or under the sun surface.

"and the distance of Jupiter from the sun is to the semidiameter of the sun in a proportion but a small matter greater, the common centre of gravity of Jupiter and the sun will fall upon a point a little without the surface of the sun. By the same argument, since the quantity of matter in the sun is to the quantity of matter in Saturn as 3021 to 1, and the distance of Saturn from the sun is to the semi-diameter of the sun in a proportion but a small matter less, the common centre of gravity of Saturn and the sun will fall upon a point a little within the surface of the sun. And, pursuing the principles of this computation, we should find that though the earth and all the planets were placed on one side of the sun, the distance of the common centre of gravity of all from the centre of the sun would scarcely amount to one diameter of the sun." (*The Principia*, 409)

That's the difference between gravity theory and vortex theory. **Gravity is a kind of direct attracting motion, and**

the common center is biased towards the heavy object. Vortex force is a kind of curved rotating motion, in which objects are separated from each other, and their equilibrium point is formed under the constraint of shells. Lagrangian points are not the common centers of gravity, they are just key marks of shell.

Gyro Effect, the Answer to the Riddle of Celestial Body Revolution and Spin

The previous discussion confirm that the aggregation force of free photons is the motive force of vortex motion, which is the cosmic force and the most powerful force in the cosmos. It can easily gather large nebulae together and twist them into red giant, black hole and star in turn. After that, the star exploded, producing small stars and planets, forming a stellar system and a galaxy.

A Galaxy is made up of all kinds of stars, and each star has its own system. How can stars, which are plasma, and planets and their moons, which are mixtures of ions and atoms, with different properties, orbits and speeds, be kept together in a system, orderly and smooth, for long periods of time? So that's going to explore the gyro effect. It may not be whole cause, but the indispensable factor of system stability.

Gyro is an ordinary toy. Its characteristic is that it can rotate and stand upright when whipping it, which is actually the embodiment of the principle of energy conversion and conservation. We beat a gyro with a stick, often pushing it to move in a straight line; While we whip it to make it stand upright and rotate. Its principles are: 1. Gyro is a conical object with a pointed fulcrum at the bottom, which has the function of transforming external force into centripetal force. 2. The action of whipping is first to wrap the gyro tightly and then loosen it with the front section of the whip, and apply a rotating force to the gyro from its surface in. It can be considered as a swirling motion.

This is how the revolution and spin motion of celestial bodies are formed. Vortex motion not only makes stars and planets into spheres, which is a more effective form to transform centrifugal force into centripetal force than a gyro, but also implants angular momentum on stars and planets like whips, forcing it to rotate. Atmospheric pressure, gravity and ground friction will slow down the gyro speed, and if we don't whip it again, it will fall down. **Without these resistances, stars and planets can spin and revolve forever after gaining angular momentum from cosmic force and vortex force.**

Spinning gyro also produces centrifugal force, but it is a solid, so centrifugal force can't work, which, following the principle of energy conversion and conservation, is converted into centripetal force, that is, spinning motion. If it is jelly, the centrifugal force will act immediately, and the gyro will disintegrate before it turns enough. If we wrap it with a soft membrane, it will not disintegrate. Part of its centrifugal force is converted into centripetal force, which makes the gyro rotate, while part of the centrifugal force is converted into pressure, which makes the gyro become a flat disk. If we add a shell to it at this time, such as put it in a closed gaseous box, it will rotate and become flat, but the flat rate is much smaller than when it is outside the box. This is the case of stars with shells, and the sun is a typical one, which is almost an ideal sphere.

There is a 'spinning bowls of water' item in Chinese acrobatics. The performer ties two bowls full of water to each end of a rope, then holds the rope separately with both hands and rotates it so that the two bowls gradually rise from the vertical to the ground to the horizontal state, or even slightly over 180 degrees to the ground, while the water remains in the bowl without any loss. This is also an example of a blocked centrifugal force.

The item shows that this blocked centrifugal force can also push objects into a space with lower pressure, that is, allowing two bowls raising.

Inference. **Any object with a high enough spin speed will be able to lift itself off the ground.** That's why tornadoes can push ground objects up into the air (a storm can't, because it doesn't spin, it just blows things around) and why gyroscopes stand upright as they spin.

The reason why stars and planets can run stably is the same, except that they will not 'rise up' due to the pressure in the space around them is the same, actually no pressure. Their centripetal force, which is transformed from centrifugal force, is partly transformed into contraction force and partly into spin speed.

Gyro motion has another characteristic, **its axis can turn freely in one or more directions and thereby maintain its orientation regardless of any movement of the base.** In the spiral arms and in the Milky Way, the solar system makes complex movements around the arm axis and the center of the galaxy, and the direction is constantly changing. Yet, the sun and planets can always be stable in a plane, and the earth and moon can stably orbit each other in their system, which is the reason.

Newton claimed that planets and other objects were inherently "be of rest, or of moving uniformly forward in a right line" (*The Principia*, 3), and that centripetal force was the force that "planets are perpetually drawn aside from the rectilinear motions, which otherwise they would pursue, and made to revolve in curvilinear orbits." (Ibid., 3)

For CFVT, **like gravity, centripetal force is not an inherent force of the celestial body, but is converted from centrifugal force.** Celestial bodies gain angular momentum from vortex forces, which must produce curvilinear motion. Curved motion, especially rotary motion, produces centrifugal force. Stars and planets do not disintegrate in rotational motion, mainly due to the action of the shell. It is like a whip to a gyro, keeping the stars and planets sound and moving in their own obits stably. **Centrifugal force is the definite product of all rotating motions, whereas centripetal force is the merits of the shell and vortex motion, the gyro effect makes these two opposing forces coordinate.**

"The reason why matter eavenly scattered through a finite space would convene in the midst you conceive the same with me; but that there should be a Central particle so accurately placed in the middle as to be always equally attracted on all sides & thereby continue without motion, seems to me a supposition fully as hard as to make the sharpest needle stand upright on its point upon a looking-glass."

Above is a quote from Newton's 1693 reply to a letter to Pastor Richard Bentley, expressing his amazement and confusion of the sky. I think the gyro effect is pretty helpful either.

From the gyroscopic effect, we can imagine the universe as a large hollow gyro with numerous medium-sized rotating gyros (galaxies), that with a large number of smaller gyros (stellar systems) inside each... Down to the proton, and inside it is a tiny gyro. So layer by layer, all in spiral motion, from high speed to low speed, orderly, stably, like **a complex perpetual motion machine,** through black holes to get renewed.

What Would Happen if the Sun Disappeared?

The answer, according to Newton's law, must be that Earth will leave its orbit around the Sun about eight minutes after the Sun disappears. Like other planets, it will fly away from the Solar System under the action of tangential forces, just like the flying chair that is spinning rapidly and suddenly breaks its cable. This is also the view of most people. This book offers the opposite answer: instead of dispersing, the planets cluster.

In the previous illustration of a motorcycle acrobatics show, we saw that schism will not happen for the shell structure of the solar system is not damaged—The corona disappears but **the Oort Cloud and the Kuiper Belt remain intact,** — and the aggregated force increases.

Sunlight & the solar wind is the most powerful antigravity in the solar system. **If the Sun suddenly disappears, then this antigravity disappears with it. The stable orbit movement of the planets will lose support at this time, and the movement of the then solar system will turn into a rapid centripetal contraction movement.** Under the action of vortex force and the angular momentum of the planet, the overall volume shrinks, and the rotation speed increases. In the original position of the sun, a vortex centre and a vortex black hole forms in succession. Then, the ignition of the black hole causes it to become a small sun — only 2‰ of the mass of the original Sun, not even a speck of dust will leave the solar system. And it is possible to exceed this ratio by dragging in some of the Oort cloud and Kuiper belt stuffs — and spin at a high speed. This small star will go through what the early sun went through: it will shrink, explode (blast out planets), expand, shrink, and explode again. Or, as Jupiter does, it will intermittently glow and explode to produce moons & rings. If you think of the Jovian planets as small stars rather than large planets, as my book suggests, the formation of this small sun is not surprising.

Astronomical Facts Opposing the Big Bang-Cosmic Inflation Theory

That nature exists, it would be absurd to try to prove; for it is obvious that there are many things of this kind, and to prove what is obvious by what is not is the mark of a man who is unable to distinguish what is self-evident from what is not. (Aristotle. *Physics*, BOOK II, Part 1)

The astronomical facts supporting the Big Bang-Cosmic Inflation theory are controversial and uncertain. The following astronomical facts opposing this theory are unquestionable and definite.

First, **all galaxies & star systems are vortex structure. Vortex motion is characterized by centripetal convergence rather than outward diffusion, so that no object — from dusts to giant stars — can escape their own vortex.**

Second, the solar system has no tendency to spread. Stephen Hawking mentions in his book that the movement of Earth in its orbit around the Sun produces gravitational waves. The effect of the energy loss will be to change the orbit of Earth so that it gradually gets nearer and nearer to the Sun, eventually collides with it, and then settles down to a stationary state. (2016, p.102) **If there is a trend of diffusion in the solar system, Earth's temperature will show a trend of continuous decline. In fact, the solar system does not have a contraction trend, otherwise Earth's temperature would show a trend of rising all the time, no any ice age.**

Third, the Milky Way has no diffusion trend. This book points out that all vortex motions are characterized by the phenomenon that rotational speed increases step by step from outside in, and accordingly matter and energy are concentrated at the centre. And the reverse is true for mechanical rotary motion. Astronomical observation results show that the existence of the Milky Way is caused by the former mode.

Fourth, there is no diffusion of galaxy clusters. On the contrary, **the phenomenon of galaxies converging and colliding has been predicted many times by astronomers.** US astrophysics professor Fulvio Melia pointed out through computer simulation that Andromeda had already swallowed a smaller galaxy about thirty million years ago. In his book *The Birth of Stars and Planets*, John Bally, another American astronomy professor, predicted that the Milky Way would collide with the large and small Magellanic Clouds in the next billion years, and with the Andromeda galaxy three billion years after this. (2006, p.228) When accounting for these facts, the universe expansion theory is also the default. But astrophysicists argue that these are all phenomena of matter and that matter accounts for only 4–5% of the total energy — mass is not enough to stop the expansion of the universe.

However, general relativity tells us that matter, space, and time are three in one. There is matter and so there is time and space; if there is no matter, then there is no time and space. The only universe we're talking about is the physical universe in which we live. Without matter — without humankind — there is no universe.

Of course, in addition to matter, there is energy, which is also the source of matter. This book equates cosmic energy with light and claims that the light field is a ball of light that, at twice the speed of light, diverges outward into space. However, as mentioned earlier, pure energy (light) does not occupy space. This is not expansion of the universe, nor is it expansion at all. No one would declare a light bulb to be expanding just because it was glowing.

"In Friedmann's model, all the galaxies are moving directly away from each other. The situation is rather like a balloon with a number of spots painted on it being steadily blown up. As the balloon expands, the distance between any two spots increases, but there is no spot that can be said to be the center of the expansion. Moreover, the farther apart the spots are, the faster they will be moving apart. Similarly, in Friedmann's model the speed at which any two galaxies are moving apart is proportional to the distance between them. So it predicted that the red shift of a galaxy should be directly proportional to its distance from us, exactly as Hubble found." (Hawking, 2016, p.50)

RETORT.
According to Friedmann's model, all 'spots painted on it being steadily blown up. As the balloon expands, the distance between any two spots increases, but there is no spot that can be said to be the center of the expansion. Moreover, the farther apart the spots are, the faster they will be moving apart,' never get close to each other again, and collision is impossible. What are the causes and forces that could cause two stars, two black holes or two galaxies to collide in a space of rapid and accelerating expansion?

A New Look at the Big Bang

Explosion is one of the inevitable results of the spiral motion of a celestial body. The universe is the largest whirlpool, so its explosion must be most violent.

Explosion is a kind of diffusion movement and expansion movement, but it produces turbulence. Vortex motion can stop the diffusion movement and replace the expansion movement. This is a significant difference from standard cosmology.

The reason for this difference lies in the starting point, which in standard cosmology is the singularity where the universe emerges from nothing. The explosion that pushes the universe to expand, and the force (matter) that holds it back is so weak that it keeps the expansion from accelerating. "Hubble's observations suggested that there was a time, called the big bang, when the universe was infinitesimally small and infinitely dense. Under such conditions all the laws of science, and therefore all ability to predict the future, would breakdown." (Hawking, 2016, p.10)

CFVT assumes that after the universe becomes a maelstrom, a large number of local explosions will occur, and finally coil into a huge black hole with a diameter of about one light-day. After the black hole ignites and explodes, a new universe is created.

Vortex motion is the greatest force in the universe, acting on the loosest and most easily compressible object (nebula), and its target is a singularity. However, it is also the movement of matter, and it follows the physical laws. **In the process of contraction, including what astronomers call collapse, a series of events such as an increase in temperature, phase changing and explosions will occur, thereby stopping or reversing the contraction movement, meaning that the singular goal will never be realized.**

From the perspective of vortex theory, the so-called Big Bang is actually the highest level of celestial vortex explosion ever studied by human beings. But we know it cannot be the overall explosion of the total universe, because it is infinite. This also stipulates that vortex theory is compatible with multiverses.

The universe must shrink before it explodes. There is no guarantee, in standard cosmology, that another Big Bang will happen, and the universe could keep expanding faster and faster. If it does shrink, according to Hawking, it will be 10 billion years from now. The cosmic explosion described in this book is a series of local explosions, and the final total (of our confirmed universe) of the explosion will be a reincarnation process. However, based on the aforementioned observations in the constellation of Phoenix, I have changed my mind.

The Phoenix event suggests that black holes are much more efficient at making stars than I originally think. In the process, the light field EEEF may play a role in making the output of the black hole greater than the input, and the total mass of the new stars may be greater than the total mass of the stars that have been spun into the black hole. If this were the case for all galaxies, the Big Bang would not have occurred and would have been replaced by an endless series of smaller explosions.

There are many indications that even galaxies such as the Milky Way and the Andromeda will not have one big final explosion but will have endless cyclic explosions. In the end, we might have to go back to Einstein's equilibrium universe.

The discovery of Ton618, however, suggests another possibility. It was first observed by astronomers in 1957 and, after more than 60 years of observations and calculations, was identified as a giant quasar, one of the brightest objects in the known Universe. It is located near the North Galactic Pole in the constellation Canes Venatici, and contains the largest known black holes, with a mass of 66 billion M$\odot$ (solar mass), and spins at about 7,000 km/s.
(https://www.space.com/black-holes-can-reach-

stupendously- large-sizes.html) If this data and information is true, according to this book, it is a super-galactic black hole that is absorbing the surrounding galaxies, and the Milky Way is sure to be among them.

In the equilibrium universe, if the giant vortex in which Ton618 is located is an independent system, and there is no higher-level system on it, or it is not subordinate to another system, we can recognize it as the universe in our concept, ignoring countless other universes coexisting. Over a long period of time, say 1 trillion years, all of the matter in this universe will be rolled into Ton618 and begin the next cycle.

The Ton618 fact proves that a black hole is not bottomless, its capacity is limited, and it cannot absorb all the swirling matter. It is likely to have a critical value at which, like a smelting furnace, it will spew out the molten material to form new celestial bodies, and singularity can never occur.

There is a consensus in the scientific community that a black hole evolved from a giant star, and its volume is determined by the mass of the giant star. By analogy, the volume of a cosmic black hole is determined by all the mass of the universe, and it must be the largest black hole. How can it become a singularity without any mass?

The Crisis of Quantum Physics

I think I can safely say that nobody understands quantum mechanics.
Richard Feynman

"Rule Three. Concerning objects proposed for study, we ought to investigate what we can clearly and evidently intuit or deduce with certainty, and not what other people have thought or what we ourselves conjecture. For knowledge can be attained in no other way.

Rule Eight. If in the series of things to be examined we come across something which our intellect is unable to intuit sufficiently well, we must stop at that point, and refrain from the superfluous task of examining the remaining items."

(Descartes, 1985, p.13, p. 28)

Quantum theory is considered to be the most successful theory in the history of science that can stand the test of experimentation. However, from the day it was born, it has been in crisis.

Firstly, quantum physics has no birth right. According to **Descartes' principle, the subject and theme of any innovation theory must be clear.** However, up to now, the basic concepts and principles of quantum theory have not been established, and its image is blurred — Einstein called it 'spooky action-at-a-distance' in the physical world. Science pursues clarity and accuracy. Can a discipline be built on such a 'spook'? Therefore, I call quantum theory a beautiful castle in the air. It is very striking and admirable, but it lacks a solid foundation and is crumbling.

Secondly, quantum physics has no status. Scientists have established theories such as particle theory, atomic theory, molecular theory, and cytology according to the natural level of matter. Quantum science should undoubtedly be classified as particle theory. The two have many similarities and differences, but quantum theory cannot replace particle theory, let alone atomic theory. **In physic world, atoms and particles are everywhere. There is no place for quantum, it has no material foundation.**

Some argue that a quantum is an idea, an understanding, and not a particular particle. This defence is weak. Even the most abstract ideas have a material foundation. Gravitons, for example, though not yet discovered, have a definite physical basis. Another example is a UFO. The concept itself is not clear, but its material foundation is very clear; that is, those flying objects that have been seen but we have failed to explain how or why. Hawking claimed that a 'photon is a quantum of light' (2016, p.241), and that raises the question; photons and light come from all luminous objects, where does a 'light quantum' come from?

Finally, the task of quantum physics is not clear. In the beginning of 1900, British physicist Lord Kelvin gave a speech entitled 'Nineteenth-Century Clouds Over the Dynamical Theory of Heat and Light' in which he claimed that in physics, there were only two remaining little problems: the theory of ether and the theoretical explanation of blackbody radiation. It was the discussion and study brought about by these two little clouds that motivated the two most important fields of physics research in the twentieth century: relativity and quantum mechanics.

It goes without saying that solving these two major problems is the main task of quantum theory. Unfortunately, in this respect, only some mathematical formulas have been provided, but these have not been solved in theory. For example, after the Michelson-Morley experiment denied the ether, the problem of the medium through which light, electricity, magnetism, and gravity propagate, remained unresolved. Scientists still default to the action at distance or use Hawking's virtual photons.

Unable to find a place in the physical world, some people turn to the spiritual world. On the internet, we see that the first exponent of this theory was no one else but Max Planck himself, the founder of quantum theory. A statement he made in 1944 makes it clear: matter is consciousness.

"As a man who has devoted his whole life to the most clear headed science, to the study of matter, I can tell you as a result of my research about atoms this much: There is no matter as such. All matter originates and exists only by virtue of a force which brings the particle of an atom to vibration and holds this most minute solar system of the atom together. We must assume behind this force the existence of a conscious and intelligent mind. This mind is the matrix of all matter.
I regard consciousness as fundamental. I regard matter as derivative from consciousness. We cannot get behind consciousness. Everything that we talk about, everything that we regard as existing, postulates consciousness." (*The New Science*, 1959)

After him, there are many followers and innovators, with quantum entanglement, quantum superposition, quantum woo, and other phenomena, they tried to construct quantum consciousness— quantum soul theory. Some quantum scientists have gone further, proposing theories such as consciousness creates reality, observer creates reality, and so on.

Quantum physicists discovered that physical atoms are made up of vortexes of energy that are constantly spinning and vibrating, each one radiating its own unique energy signature. Therefore, if we really want to observe ourselves and find out what we are, we will find that **we are beings of energy and vibration, radiating our own unique energy signature.** This is fact and is what quantum physics has shown us time and time again.

On this basis, they believe that matter does not exist, that nothing is solid, and everything is energy. Richard Conn Henry, Professor of Physics and Astronomy at Johns Hopkins University, called on people in his article *The Mental Universe*, "The only reality is mind and observations, but observations are not of things. To see the

Universe as it really is, we must abandon our tendency to conceptualize observations as things... **The universe is immaterial — mental and spiritual. Live, and enjoy."** (*Nature*. Vol. 436, p.29, July, 2005)

In my opinion, matter is imprisoned energy, but I don't deny that matter, energy can't supplants matter. Only for the reason that energy is invisible and matter is visible, it is convenient for observation and measurement. Even if I want to observe energy, it can only be achieved by material means. Consciousness (and spirit) can't supplant matter either, because it belongs only to men, not to things. Giving consciousness to particles or quantum directly or in disguised form is a popular practice in present, starting from Hegel's time.

"German physicists for some time accepted this (Kantian) **pure dynamic.** But in spite of this, the majority of these physicists in modern times have found it more convenient to return to the Atomic point of view, and in spite of the warnings of Kästner, one of their member, have begun to regard Matter as consisting of infinitesimally small particles, termed 'atoms' which atoms have then to be brought into relation with one another by the play of forces attaching to them— attractive, repulsive, or whatever they may be. This too is metaphysics; and metaphysics which, for its **utter unintelligence**, there would be sufficient reason to guard against." (*Shorter Logic*, §98)

Humans have been paying attention to atoms for 2,500 years. Today we have a profound understanding of their types, structures and properties, and we apply this knowledge to make a large number of electronic devices and nuclear bombs. But the atom is still a particle that we know most thoroughly embodies human wisdom, and it itself is still 'utter unintelligence', and we can never expect it to tell us any truth.

Seen from this perspective, particle physics is a fairy tale.

There, **particles and their agents (messenger particles) seem to possess the magical four forces and 'utter intelligence'.** They know how to precisely control their position and motion, including attractive & repulsive; the orbital motion within the atom and the orbital motion of celestial bodies in sky, their entanglements and annihilation, to achieve a stable harmony of the universe.

The Essence and Prospect of Quantum Theory

What is a quantum? There is a clear description in Rodney A. Brooks' book:

"In the centennial year of 1900, Max Planck introduced the idea that **the EM field is not a continuous 'classical field', but is made of pieces, or chunks, that he dubbed quanta** (from Latin quantum meaning 'how much'). While Maxwell's classical EM field can be arbitrarily small, quantum fields are made of chunks that cannot be further reduced. Quanta may overlap each other, but each one maintains its separate identity; it lives a life and dies a death of its own. In that sense, and in that sense only, field quanta resemble particles." (2016, p.20)

Einstein opposed quantum theory all his life but, ironically, he made a significant contribution to quantum theory. Quantum theory was improved by his participation, and he is even considered as one of the founders of quantum theory. Brooks commented, "While Einstein may not be the father of QM, he surely is its grandfather." (Ibid., p.15)

The greatest feature of quantum theory is that it cuts energy — light, electricity, magnetic energy, and other material energy — into chunks, splits continuous waves into pieces of 'chunks', and changes determined atomic behaviour into the drift probability of 'chunks', which is the corpuscular way to deal with holism. This theory deals specifically with pure energy, and not only physical energy, but human consciousness as well.

Corpuscular theory cannot deal with the whole, but it cannot completely avoid the whole subjects, as pure energy is the prominent one. Material energy cannot be separated from a material carrier, which can be treated separately. Pure energy is a whole, and corpuscular theory cannot deal with it but has to face it, so the emergency measure is to quantize it — mathematically splitting it into chunks.

I classify atomism, particle theory, gene theory, etc., as being parts of corpuscularianism. Quantum is not as distinctly corpuscular as these other theories, and so I call it the shadow of the corpuscle. It is impossible for scientists to use quantum theory to solve the problems that the other corpuscular theories have failed to solve. Einstein has a good saying here, **"The significant problems we face cannot be solved at the same level of thinking we were at when we created them."** (The New Quotable Einstein, 2005)

From Brooks' book, we know that the new task of quantum theory is to solve the puzzle of wave-particle duality. The protagonist of corpuscularianism is the particle, and the field is its attribute, which produces a large number of different fields, such as the strong field, weak field, electromagnetic field, gravitational field, etc. (more than 10 kinds of fields are listed in Brooks' book), and even every particle or electron can create its own field.

Quantum theory turns the subject upside down and makes the field the leading role. This theory, in its true sense of 'no particles, only fields' (p.13). The field inundates all particles, and that problem of wave-particle duality is digested. "Einstein came to believe that reality must consist of fields and only fields." (p.17) Brooks quoted Einstein's following passage.

"What appears certain to me, however, is that in the foundations of any consistent field theory, **there shall not be, in addition to the concept of field, any concept concerning particles.** The whole theory must be based solely on partial differential equations and their... solutions. — A.Einstein. *The theory of relativity, and other essays*, p.34." (Brooks, 2016, p.17)

However, field quanta maintain their separate identity, and resemble particles. They are also contradictory to the consistent field theory. Einstein's dream of establishing a unified field also failed.

Predictably, if my book continues to be ignored, quantum theory will have a period of development before better energy theory & field theory (but I believe my energy theory and field theory can't be bypassed) come into being and make it widely accepted. The birth of a new theory is mostly difficult and tortuous, there is no smooth road to travel. That researchers of energy can use QM as the leading theory is a lucky thing.

Quantum Theory May as Well Be Replaced by Photon Theory

The first reason is that the quantum has no material basis, whereas the photon has, namely, all luminescent substances. Photon is also a kind of quantum, called **light quantum**. It was not put forward by Max Planck, the founder of quantum theory, but by Einstein, the opponent of quantum theory.

The second reason is that quantum theory and photon theory are **similar in character.** The quantum is one chunk after another, whereas the photon is one wave after another. Both have wave–particle duality, but the two properties are more specific to the photon than to the quantum. In addition, by observing sound waves and water waves, we come to the conclusion that any wave is a way for transmitting energy, which is continuous and forms consistent fields. Energy will not be attenuated but will be enhanced in the transmission process. We have determined that light propagates in the form of waves in light field, and no medium is needed, while the way of quantum propagation is unclear — is it relying on virtual photons? Hawking thought "quantum is an indivisible unit in which waves may be emitted or absorbed." (2016, p.241) The holistic view is that waves must be continuous, that is, they exist just in field. Isolated wave only exists in geometric graphs.

The third reason is that photon theory clearly describes antiparticles and thus negates the concept of antimatter. Photons come from a light source and radiate outward in the form of light balls. In other words, each photon has a photon with an opposite direction of motion and spin, and the two are symmetric one by one, or even identical and unified, and we can determine this with our intuition and **thus negate the concept of antimatter.** Although quantum theory predicts the existence of antiparticles, this is only necessary for mathematical calculation. There is no evidence in reality.

The fourth reason is that photon theory and light wave holography theory can clearly describe the mirror image of the light field, **thus negating the particle entanglement hypothesis.** In a solar flare, for example, the radiation is fan-symmetric, so it is not surprising that there is synchronization between any pairs of two points at the far end. There is no mystery here and no entanglement.

All particles come from photons, and light is energy. The darkened light theory, the photon theory, the light ball theory, the light field theory, and so on, synthesize the new energy theory.

No Messenger Particles

"The indestructibility of motion cannot be merely quantitative, it must also be conceived qualitatively; matter whose purely mechanical change of place includes indeed the possibility under favourable conditions of being transformed into heat, electricity, chemical action, or life, but **which is not capable of producing these conditions from out of itself, such matter has forfeited motion;** motion which has lost the capacity of being transformed into the various forms appropriate to it may indeed still have dynamis but no longer energeia, and so has become partially destroyed. Both, however, are unthinkable." (Engels, *Dialectics of Nature*)

We buy tickets to watch the boxing match in view of the famous boxers. If what we saw at the scene was not the boxer's fight, but their doubles or their gloves were fighting, we would quit. **The messenger particles are the surrogates or shadows of the elementary particles.**
Messenger particles are also called intermediate particles, exchange particles, force-carrying particles, medium particles, or bosons. Their task is to eliminate action at a distance and to maintain interactions between particles. The law of universal gravitation cannot be short of the action of distance; Newton admitted that he could not explain this phenomenon, and interrupt physics research. Later scientists intended to make a breakthrough on this, and various messenger particles came into being. They carry various forces, like messengers, allowing particles to acquire forces, connect with each other, and interact with each other, even in long distance.
The first medium that people thought of was ether, which came from ancient Greece. Greeks believed ether was the fifth and highest element after air and earth and fire and water; was believed to be the substance composing all heavenly bodies. Two thousand years later, René Descartes

introduced the idea of the ether into his vortex theory, and James Maxwell later applied it to his electromagnetism. Physicists at that time generally believed that ether was an all-pervading, infinitely elastic, massless medium formerly postulated as the medium of propagation of light and electromagnetic waves. The existence of ether was disproved by the American physicists Albert Michelson and Edward Morley in 1887. (This book affirms the light ether.)

Modern physicists believe that any interaction between two objects occurs through the exchange of subatomic particles. The larger the mass of the exchanged particles, the smaller the corresponding range of action.

At present, four messenger particles are popular, which act on four interaction forces of nature respectively. The messengers of a strong interaction are called gluons because they glue the quarks together into particles, like protons and neutrons, and in turn glue protons and neutrons together into atomic nuclei. Photons (virtual photons) are the messengers of the electromagnetic interaction, and particles called W and Z carry the weak force. Gravitons are believed to carry gravity, but so far they have not been found. One of the things that determines the range of an interaction is the mass of the messenger particles.

The God particle is the most powerful and sensational of these. Its scientific name is Higgs boson; It is said to be the particle that gives all other particles mass but has no mass of itself—Its spin is zero. It is smaller than all other particles but also heavier in atomic weight— In *Dreams of a Final Theory*, Steven Weinberg claims the Higgs particle is 50 times more massive than the proton. The Higgs field is an invisible energy field that fills space. On July 4, 2012, CERN announced that it had found this particle in existence.

Electrons repel each other, and their interaction also requires messenger particles.

"The force-carrying particles exchanged between mass particles are said to be virtual particles... For example, **the electric repulsive force between two electrons is due to the exchange of virtual photons,** which can never be directly detected." (Hawking, 2016, p.78)

As we know, all atoms except hydrogen have multiple electrons, so they need a large number of virtual photons to exert constant force in order to prevent the electrons from encroaching on each other. Therefore, virtual photon is the richest messenger particle.

This raises multiple questions; **where does the virtual photon come from? How do they know how many electrons in an atom need to repel each other? What interaction do they have with gluons in the nucleon?**

One of the key terms in this book, vortex force, is a new term in contemporary physics and has not yet been endorsed by scientists. A close term in physics is tidal force, which is characterized as secondary effects of gravity. It is mainly used to explain the tidal phenomena on earth, as explaining the motion of a celestial body's vortex is obviously not appropriate, and because gravity is the weakest of the four forces, and vortex is the most powerful force in the universe. In addition, their motion forms are also obviously different. Newton's law shows that gravity is the linear interaction force between objects, while vortex motion is generally an independent spiral motion. According to common knowledge, when objects move in a spiral way, no messenger particles are needed.

Another common term used in this book is centrifugal force, but it is rarely used in classical physics, and for one reason: its messenger particles are not found. Centrifugal force is the opposite of centripetal force. The latter is gravity and its messenger is the undiscovered graviton, which cannot act as the messenger of centrifugal force at the same time, and more importantly, gravity cannot tolerate antigravity. Physicists believe that centripetal force is real and centrifugal force is virtual.

Messenger particles are tools of corpuscularianism. Since the particles are discrete, independent, and powerless, communication between them requires messengers. **The**

four forces are independent, can't travel, can't cross space, and need messengers also.

From Engels's above words, **we see that one consequence of the introduction of messenger particles into physics is that it reduces the status of the protagonists and their lives are destroyed.** Quarks and electrons are forced to stick together by gluons. Gravity and electromagnetism are weak in themselves and require the services of gravitons, photons and virtual photons. The decaying atoms are so weak that the W and Z particles need to be applied to send their decayed ones away. This situation makes us even more baffled about gravity. **Since all elementary particles repel each other, matter cannot be the source of gravity, and gravity cannot be proportional to mass. Matter wildly needs an input gravitational force, such as quarks need gluons all the time; it cannot have any output gravitational force. Where does gravity come from?**

This kind of tool is unnecessary in CFVT because it is a holistic theory. In this theory, the energy of the particle itself never fails, and there is no need for any intermediary particle, and the basic state of particles is fields. The field is continuous, filling the whole space — everything and energy are all in it, connected with each other — so there is no problem of action at a distance.

The Duality and Three Conditions of Things

The existence of everything requires three conditions: a name, time, and space — of which a name is the most important. This is because **everything has duality; that is, nature and sociality.** Time and space are the attribute of things; they are not endowed by society but are inherent. The name of things is given by humankind; the sociality of things is first embodied in naming. That is to say, things are the products of humankind's thought. Although nature is the origin of all things, these things are only natural objects, not social objects, and are Kant's 'Das Ding an sich', thing-in-itself, whom Hegel remarked: "Hence one can only read with surprise the perpetual remark that we do not know the Thing-in-itself. On the contrary there is nothing we can know so easily." (Shoter Logic, §44) They are not yet real things until they have been reviewed and dubbed by Man.

The first power God granted to the first man he created was to give all living creatures names, "and whatever name he gave to any living thing, that was its name." (Bible, Gen 2:19) The Italian thinker Giambattista Vico pointed out in *New Science* that to name a thing is to define it. Naming is the master's privilege to Nature, Nature thus acquires its 'legal identity' in our world.

People name things on the basis of common sense. Names do not fully reflect the essence, but they are the first step toward determining the essence of things, an indispensable step. A good name will roughly reflect the essence. With an essential understanding of things, we can determine our attitude toward things and discover how they play their roles.

Not only do specific things need to be named independently, but they should also be classified and named according to their nature and hierarchy. There are higher or more detailed classifications above the classifications, resulting in a dendritic structure system of names, analogous to the classification and naming of books in a library.

Einstein said: "I think that a particle must have a separate reality independent of the measurements. That is an electron has spin, location and so forth even when it is not being measured. I like to think that the moon is there even if I am not looking at it." (GIGA Quotes)

His belief comes from the social attribute of things. **We cannot see particles or electrons, nor can we see the moon at any time, nor can we see most foreigners and historical figures, but we believe in the existence of these things and people because many persons have witnessed their existence, and society recognizes their existence and names them,** documents them, or includes them in textbooks. "Everything that exists stands in correlation, and this correlation is the veritable nature of every existence." (*Shorter Logic*, §135) The existence of Earth's most conspicuous satellite was verified and confirmed thousands of years ago when it was dubbed the Moon (or whatever it was in different scripts), such was its social attribute. After that, even those who have never seen the Moon could safely believe in its existence, and anyone who denies its existence is futile — often regarded as ignorant or an idiot. There is also human experience and reason behind it. I have seen the moon since childhood, and my schooling has taught me the cause of its movement, so that even if I do not look at it now, reason will convince me that it is where it ought to be. It is odd that someone doubts the authenticity of the moon, their ancestors or other things because they can't see them now.

In *The Plato Code*, I put forward a proposition about the relationship between knowledge and truth. **Truth comes from facts, knowledge comes from truth, which is the literal description and conceptual form of truth. Truth and knowledge are public property. Every truth and knowledge is correlated to a sociocultural system.**

The task of scientific research is to constantly discover the truth and enrich the treasury of knowledge. Accordingly, the task of scientists or vulgarians is to discover. After making new discovery —a new theory, a new law, or, a new star, a meteorite, a new particle, a new species of plants or animals, and a piece of UFO —the first thing they should do is name and describe their discoveries, and submit them to society in some form (books, papers, reports, dictation, photos, etc.) After being recognized by society, and archived, their findings become an integral part of social culture for public use. Thus, ordinary people don't have to watch or measure them to determine the existence of these new things, and they can use them freely as certain knowledge. Through these discoveries, society becomes strong and advanced, and the discoverer gains honor and rewards from society.

On the contrary, if someone makes a new discovery but keeps it a secret, then their discovery is not social and the public does not benefit from it, it is not public property, it is just his private property, and he will not get social honor or reward from it.

Conclusion: Things are not isolated due to natural existence but are system of social knowledge. They are named and defined by people, and gain their 'legal identity', then they get real existence. Its social attribute is the first, and its natural attribute is the object of scientific research, which is also endowed by society.

Aristotle said that man is a social animal, and this rule is especially important to the individual. After a person is born, the first thing the parents should do is to name it, register it in the archives, let it become a member of the society. A person without a name cannot enjoy all kinds of social rights and benefits and is not a real citizen, or a nonexistence.

Naming makes sense not only for individual, but also for nation and country. From the perspective of scientific research, human itself is also the object of scientific research. The first thing is to define human. So far, there has not been a unified answer, and perhaps never will be. Because this is first and foremost privilege of every individual, every nation and every country, **people basically live their lives according to their own definitions,** viz., their names, resulting in a multitude of different religions, groups, laws, cultures and worldviews. Even the United Nations cannot prescribes a comprehensive definition of man and entails all people to live by it.

We can boil these subjects down to the relationship between man and nature or the relationship between mind and nature. Hegel says that nature is as the system of unconscious thought, therefore, **"Nature in like manner is not without mind, nor mind without nature."** (*Shorter Logic*, §119) If we have this consensus, his following paragraph will be understandable.

"If thought is the constitutive substance of external things, it is also the universal substance of what is spiritual. In all human perception thought is present; so too thought is the universal in all the acts of conception and recollection; in short, in every mental activity, in willing, wishing, and the like. All these faculties are only further specialisations of thought. When it is presented in this light, thought has a different part to play from what it has if we speak of a faculty of thought, one among a crowd of other faculties, such as perception, conception, and will, with which it stands on the same level. When it is seen to be the true universal of all that nature and mind contain, it extends its scope far beyond all these, and becomes the basis of everything." (Ibid., § 24)

The following is a more difficult and profound passage, typical of Hegelianism. It means: the concept (notion) is first, things depend on it to exist. That's why everything needs one or more names first to exist.

"In other words, the notion is what is mediated through itself and with itself. It is a mistake to imagine that the objects which form the content of our mental ideas come first and that our subjective agency then supervenes, and by the aforesaid operation of abstraction, and by colligating the points possessed in common by the objects, frames notions of them. **Rather the notion is the genuine first; and things are what they are through the action of the notion, immanent in them, and revealing itself in them.**" (*Shorter Logic*, § 163)

Matter Is Atom

Matter has not yet been definitively defined. For the sake of science, I've forced it to be atomic because matter comes in the form of particles and atoms, which make up molecules and objects. My reasons for choosing atoms over particles in defining matter are as follows:

1. To obtain the individual name and determine the nature. There are 110 kinds of atoms, each of which has a definite name.
2. Stability. Compared with particles, atoms have a narrow range of motion and undergo few changes.
3. Systematization. Atoms consist of closely related protons, neutrons, and electrons that make up the tiniest and most fundamental system in the universe.
4. Accuracy. Atoms are distinguished from each other by the amount of charge, which is accurate to the extent of a pair of positive and negative charges.
5. Particularity or absoluteness. The nature, structure, and mode of motion of each atom is different and eternal, thus marking its independent status.

The foregoing five properties of the atom are not completely possessed by particles, which are the unity of energy and mass, and do not completely belong to matter. Therefore, the atom is the basic unit of the material world. Thus, comes the definition of matter:

Matter is the most basic and stable element that makes up all things in nature. Matter is atom.

This definition does not directly reveal the nature of matter, but it does indirectly reveal the nature of the atom. As long as we know what an atom is, we know what a substance is. Atoms have all kinds of properties and characteristics of matter, such as motion, mass, gravity, and space-time conditions.

Many people believe that matter is **infinitely divisible**, which is one of the foundations of string theory. Our definition here negates this view. It clearly declares that matter is atoms and objects composed of atoms, protons are the end of matter, and it is inseparable. It may be shattered, but then it is no longer matter, but energy.

The material world has both macro and micro properties, and the concept of matter is divided into the broad sense and the narrow sense. In the narrow sense, matter refers to the atom, the main object of chemical research. Above the atom is the broad sense of matter called a body, which is the main object of physics research — its smallest unit being the molecule. The former is pure matter, finite in number. The latter consists of the former; it's made up of many different substances, infinite in number, that are very different from one another. In reality, however, the two are often used intermingled because of the vague concept of matter. Celestial bodies, objects, molecules, atoms, particles, photons, and even neutrinos, are all thought to be matter. This book also adopts popular practices in many places.

Matter Is Trapped Particles

This is a more profound definition of matter. It's a common way of energy-mass conversion described in this book. Quantum scientists call this phenomenon quantum collapse and confinement, but they have not explicitly identified quarks and electrons as energy particles.

"Quantum collapse... It is spread out, but when it transfers its energy to an atom the entire field vanishes, no matter how spread-out it is." (Brooks, 2016, p.68)

"Quarks and gluons are the basic fields that constitute the strong and baryon fields, but they do not exist in free form. This is known as the Principle of Confinement." (Ibid., p.179)

Since the establishment of particle theory, people have a deeper understanding of matter, from the atomic theory of matter to the particle theory of matter. This cannot be said to be wrong, but there is the indispensable condition that particles must stop flying.

An energy particle is naturally free. In the absence of any constraints and obstacles, it flies ceaselessly and is energy. At this point, it takes up no space, has no mass, and has no visible form. When it encounters obstacles and constraints, it immediately stops flying, enters a static state, takes up space, and becomes material.

There are two main obstacles to energy particle flight. One obstacle is a dense energy field. The distribution of energy in the field is uneven. **The dense area where the particles are present is blocked from flying, and the velocity is converted into polymerization energy.** The most common result is that three quarks are polymerized into one proton, and material is generated. The other obstacle is ready-made material. Compared with the movement of free particles, matter is a static object. Once the two meet, an interaction will occur. The particles stop flying, its velocity turns into polymeric energy, and particles polymerize with each other to form protons or, with existing substances, form new substances. Therefore, matter is also imprisoned energy.

The collection of particles forms entities such as the Sun. Although the particles move violently inside, they are in an ionic state. While outside, they are lost in the form of light radiation, but the actual loss is extremely small compared with the total amount, so we can also regard all stars as huge protons. **We can't see quarks in a proton in front of us, because they are no longer just three corpuscles that have coalesced into one tiny field.** The plasma trapped by distant stars is seen by us because, not only are they huge in number, they also glimmer. This is the significant difference between the two kinds of atoms.

The related question is: what forces will imprison the particles for a long time after they have aggregated into matter and cannot regain their freedom? The answer is one word: shells.

Shells are not foreign, they are products of polymerization. In the case of the quark, it is energy that is still energy after it has been aggregated into protons, but it cannot be preserved in the form of motion. Instead, it is preserved in the form of a strong field, and the electron is its shell. People, except for L. V. de Broglie, E. Schrödinger, and a few others, are used to thinking of an electron as a light particle, thousands of times lighter than a proton, which revolves around the nucleon at high speed. **Now I think of it as a weak field enveloping the nucleon, acting as a shell. In vortex motion, atoms come together into clumps in which the electrons are no longer independent and instead combine into a single shell that wraps all the protons —The ingenious design of the cosmic spirit, in which the weak electron field of energy wraps the strong nucleon field of energy that forms the atom and everything else, has left us all in awe.**

FURTHER DISCUSSION

"When individual atoms of copper are brought together to form a bulk metal material, something unexpected happens. **The outer electron of each copper atom leaves its parent atom. Rather than orbiting single atoms, the outer electrons all begin 'orbiting' around and among ALL the atoms in the metal.** Essentially the metal's electrons are 'jumping' from atom to atom all the time, even when there is no electric current applied. Physicists call this the 'electron sea' or 'electron gas' of the metal." (Beaty, 1995)

In the case of water, quarks trapped in proton shells, nuclear shells, atomic shells, and molecular shells are very stable. **These shells are essentially different electron fields.** When the water is cooled into ice, an additional layer of ice shell makes the water more stable, and ice can remain unchanged for hundreds of years, such as in the case of glaciers.

Primitive people knew how to drill wood to make fire. It is also a skill for modern people to survive in the wild — but where does the fire in wood come from? The answer is simple: as the pieces of wood rub against each other, cell shells, molecular shells, atomic shells, nuclear shells, atomic shells, and trace amounts of proton shells are destroyed one after another, freeing a few quarks, while wood chips and air burn under their impact. The reason for striking flint to produce sparks is the same.

The fact that matter burns and heats up makes it certain that all matter contains energy. This belief is reinforced by Einstein's theory of mass–energy automatism. However, the nature of energy is still uncertain. Most people think of energy as tiny particles smaller than protons, such as phlogistons, calories, photons, electrons, and gravitons. The new definition corrects this misconception and unites energy and particles.

Quark & its shell are energy, invisible by any means. All one can see is the interaction between quark and

matter. Bruce H. Lipton, PhD, an internationally recognized leader in bridging science and spirit, has a vivid description in his book, *The Biology of Belief: Unleashing the Power of Consciousness, Matter and Miracles*, which can help us understand the interior of the atom.

"Quantum physicists discovered that physical **atoms are made up of vortexes of energy that are constantly spinning and vibrating; each atom is like a wobbly spinning top that radiates energy.** Because each atom has its own specific energy signature (wobble), assemblies of atoms (molecules) collectively radiate their own identifying energy patterns. So, every material structure in the universe, including you and me, radiates a unique energy signature. If it were theoretically possible to observe the composition of an actual atom with a microscope, what would we see? Imagine a swirling dust devil cutting across the desert's floor. Now remove the sand and dirt from the funnel cloud. What you have left is an invisible, tornado-like vortex. A number of infinitesimally small, dust devil–like energy vortexes called quarks and photons collectively make up the structure of the atom. From far away, the atom would likely appear as a blurry sphere. As its structure came nearer to focus, the atom would become less clear and less distinct. As the surface of the atom drew near, it would disappear. You would see nothing. **In fact, as you focused through the entire structure of the atom, all you would observe is a physical void. The atom has no physical structure**—the emperor has no clothes! Remember the atomic models you studied in school, the ones with marbles and ball bearings going around like the Solar System? Let's put that picture beside the "physical" structure of the atom discovered by quantum physicists. No, there has not been a printing mistake; atoms are made out of invisible energy not tangible matter! So, in our world, material substance (matter) appears out of thin air. Kind of weird, when you think about it." (2008)

NOTE: Many things have two or more definitions, such as energy, spirit, beauty, philosophy, science, art, man, etc. Accordingly, I have offered two definitions for matter. They are interrelated and promote each other.

Two Kinds of Bodies: Ionic Bodies and Atomic Bodies

The work of classifying objects is of course premised on the definition of matter, and now that I have established two new definitions of matter, it is not superfluous to form a new classification of objects.

This is a new concept of matter, I didn't have it before, it comes from my musing about plasma. It is generally considered that plasma is an ionized gas-like substance composed of the atoms after electron stripping, and the positive and negative ions generated by ionization of the atomic group, which is electrically neutral. It is the fourth state of matter, which makes up most of the stars, and therefore accounts for more than 99% of matter in the universe. Therefore, I now elevate it to first place in the physical world. We also need a fresh look at its origins and nature.

The interchangement of energy and matter, or the interchangement of ions and atoms, is a universal phenomenon in the cosmos. Now, to put it another way, **the interchangement of plasma and atoms, is also a universal phenomenon.**

It first causes the loose nebula to contract into a vortex, turning particles into atoms, which continue to contract, splitting the atoms into ions to form a red giant then a black hole. As the black hole continue to contract, it reverts to an ionic body, then coalesces into plasma and glows, and that is a new star. According to CFT, they all are ionic resonant vibration objects, namely electricity stuffs.

A plenum of new observations suggests that black holes are not bottomless sinkholes with no exit, but intermediate-rotating bodies with a balance of entry and exit, which I call it the foetus of a star or the melting furnace of a celestial body. The former refers to stellar black holes, which are transformed into a single star under the action of vortex force. The latter refers to a galactic black hole that eats stars and spits out ionic material to form new objects.

Like matter, plasma has three states: gaseous, liquid and solid, which form the outer layer, middle layer and core of a star. Planets, which are eruptions of stars and also plasma, have all three states. **Stars explode result in most of their angular momentum transferring to the planets, so the planets are born with a strong spin force or pressure or contraction force, prompted planetary transformation, a large number of plasma objects turned into atomic objects, formed a solid crust and its various atomic stuffs,** such as air, sea, animals and plants, while its internal is still a plasma, That is the reality of our planet.

Thus, we have a cosmological proposition: **All things on earth evolved from plasma.**

Its important condition is temperature, that is, particle velocity. From scientific common sense, we know that the core of an atom is stable and immobile, but the electrons around it are always in motion and often leave the nucleus. Bare protons with all electrons removed are ions. From the beginning, the sun is a plasma composed of bare protons. They have a strong impulse to combine with electrons and reduce them to atoms. The high-speed movement of ions and frequent collisions make the sun shine.

This is also the case when the earth is just out of the sun. It only gets strong angular momentum from the sun, and it turns into strong pressure, which makes the gaseous surface plasma phase change into liquid and solid plasma, and further deepens into solid crust and the matter on it. This is the result of ions acquiring electrons and reducing them to atoms.

Among them, few were reduced incompletely, and they either failed to absorb enough electrons or absorbed too many electrons to form cations & anions on the earth respectively. In this way, we become clearer about the members of the material family: **Plasma is the undisputed parent, and atom, cations & anions are its descendants.**

Sunspots Are Windows to the Sun Hypothermic Spheres

Since plasma accounts for more than 99% of the total amount of matter, when we talk about objects now, we will take it as the most important and common object, but its basic unit needs to be changed.

An ion does not conform to the first definition of matter because it is not an atom, but the debris of an atom. However, plasma meets the second definition of matter. We can regard it as a giant nucleon of trapped protons, neutrons, but its shell (the sphere surface of a star) and contents are much larger than those of a nucleon. Let's turn to the stars now, taking the Sun as an example.

The Sun is a plasma, a giant ion ball. Its interior is stratified, like the Earth, and it has three states. The core of the Sun is its solid plasma, similar to the core of the Earth. From the core to the inner edge of the photosphere, is its liquid plasma, while the photosphere (and chromosphere) is its gaseous plasma.

This classification has an obvious problem at its core. Astronomers generally believe that the Sun's core, the densest part, is a zone of intense nuclear reaction, where four hydrogen atoms coalesce to form a helium atoms at all times, known to researchers as **'hydrogen burning'**. This is like a continuous large-scale nuclear explosion, reaching temperatures as high as 16,000,000K, but the heat generated by it is difficult to conduct out, and the Sun's core should be gaseous. However, there is no consensus on the whole idea, because it violates the proposition that stars generally have a metal core.

If we imagine the core of the Sun as a huge nuclear reactor, a solid object, continuous nuclear explosions will soon destroy it. High temperatures will vaporize all atoms, including hydrogen, helium, lithium, iron, and so on, into ions, and nuclear fusion cannot be carried out in ionic gas. From the existing astronomical observation data, I determine the dichotomy, that is, the photosphere of the Sun, is gaseous plasma and the solar body wrapped by the photosphere is liquid plasma.

Its density and pressure must increase step by step from the outside in, the core density and pressure being the highest. Here, I agree with the astronomers, but I disagree with them on the temperature. They believe that the Sun's temperature and density coincide, and that it also increases step by step from the outside in, with the highest core temperature driving nuclear fusion. **I insist on the contrary that the core temperature is the lowest, no more than 1,000K, which is the melting point of most metals, resulting in no nuclear fusion.**

There are two evidences for this inference: alcohol lamp flame and sunspots.

The temperature of the flame of the alcohol lamp drops step by step from the outside in, with the lowest core temperature, commonly known as **Wick effect**. The reason is that a flame's core is the densest and most pressurized, limiting the motion of molecules and atoms, and speeding them the least. On the surface of the flame, the opposite is true, making it the hottest area. In addition, the surface layer of flame is also the area where alcohol and air are polymerized, and combustion takes place here. Though the core of the flame was surrounded by fire, but the mist of alcohol molecules around the wick does not burn because of the lack of oxygen, and its light is radiated from the surface. That is the case of a star, except that it doesn't burn even in its surface.

Next we focus on the sunspots.

When we observe the fireballs produced by ordinary incenses and the fireballs produced by nuclear bomb explosion, we can't find spots similar to those that periodically occur on the surface of the Sun. The fireball is unified. Its temperature can increase step by step from the inside out (mostly in the former combustion) or decrease in the same way (mostly in the latter combustion), but it is impossible to produce many low temperature spots on the spherical surface in isolation, namely sunspots.

This phenomenon occurs only when the fireball is enveloping a cryogenic sphere, which is exactly the true state of the Sun. The sunspots, Galileo Galilei guessed, were clouds floating above the surface, blocking some of the sunlight from reaching us. Modern astronomical observations, however, suggest that sunspots go deeper than the photosphere.

If we reverse the position of the two, **the photosphere and chromosphere are clouds floating above the sun surface, the photosphere is dense ion cloud, similar to the earth's atmosphere, and the sun surface is a vast relatively low temperature ocean of plasma,** the puzzle will become clear. Alas! We earthlings have been bewildered by the dazzling appearance of the sun!

Comets and stars are very different in nature, but they have something in common in appearance: their relatively cold bodies are surrounded by bright, hot clouds. Here, a phrase about the brightness of comet head in Newton's *The Principia* can be used as a reference.

"Thus the earth, if it was viewed from the planets, would, without all doubt, **shine by the light of its clouds, and the solid body would scarcely appear through the surrounding clouds.** Thus also the belts of Jupiter are formed in the clouds of that planet, for they change their position one to another, and the solid body of Jupiter is hardly to be seen through them; and much more must the bodies of comets be hid under their atmospheres, which are both deeper and thicker." (485)

After Newton, a German deserter William Herschel, who later became the first president of the British Royal Astronomical Society, with his sharp eyes, saw the true face of the sun through its thick clouds. "Who suggested that **the sun was a dark, cold body with a flaming layer of gases all about it. The sunspots, by this view, were holes**

through which the cold body below could be seen. Herschel speculated that the cold body might even be inhabited by living beings." (Asimov, 1993)

Regretfully, few people have taken this idea seriously, and many have focused on a related one: The variations in sunspots would correspond to variations in solar irradiance which might affect the heating of the Earth's atmosphere and the Earth's weather.

The sun's photosphere is bright, and it wraps up a huge plasma ball with a liquid surface and a solid core that's glowing red. **Sunspots are windows opened by the 'ionspout' of the solar plasma ocean, or eyes of solar storms, allowing earthlings to peer into its interior.** Ionspouts are cataclysmal motion of plasma in the same direction, which create sunspot magnetic fields as high as 1,000-3,000 Gauss. Ocean currents in the Earth's southern and northern hemispheres are in opposite directions, and the plasma ocean currents in the sun's two hemispheres obey the same laws of physics, going in opposite directions, which is why scientists have been able to detect large numbers of spots of opposite polarity appearing in pairs on the sun's equator belt.

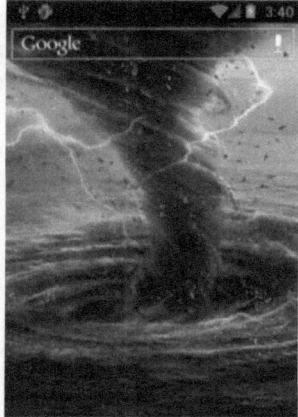

Hotline public picture

A star is a plasma whose temperature is directly proportional to the velocity of ions and inversely

proportional to the cube of its density. Its core is the densest, where ions move in a tight space near a standstill and thus have the lowest temperature. It has the lowest surface density, where ions travel in a wide space, and thus the higher speed as well as higher temperature, and a star shines from its surface, the photosphere. The space above is much wider, and ions can travel at the speed of light, resulting in the high temperature of millions of degrees Celsius in the solar corona. But its density has also fallen sharply, so the background temperature remains low.

NOTE: it seems to contradict the previous description that the temperature of the vortex of celestial bodies rises gradually from the outside in, and the core temperature is the highest.

They are all vortex motions, the early vortex of the star and its late vortex, so we need to see the difference between the two motions and the final result. Let's denote these two motions with E and L respectively.

First, the material composition. **E is a mixture of particles, atoms, molecules, chunks, etc., with most of their energy locked up. L is pure plasma, its energy is in a semi-trapped state.**

Second, the volume. E, a loose, bulky object whose particles move at high speeds. L, a relatively small and solid object whose particles are limited in movement and slow in velocity.

Third, the results. E experiences mass-energy conversion. Under the contraction pressure, all kinds of material shells are crushed and material energy is released, resulting in high-speed motion. L experiences energy-mass conversion. Under the contraction pressure, ion motion decreases step by step from outside in, the polymerization movement increases and the material shell forms, resulting in the energy of the plasma core being completely trapped and evolving into an atomic body.

Simply put, E is ignition and L is flameout.

How Can Plasma Be Preserved in Stars permanently?

Quarks form protons and remain steadily in the nucleon, so how can plasma remain steadily in the stars? I saw on Dr David Stern's website, *Stargaze*, that a nine-year-old girl phrased the question differently; she asked, how do make a container that can hold plasma?

Stern thought it a tough question. A plasma is a very hot gas, hot enough to conduct electricity, so it better not touch the walls around it! If it does, it will pass some of its heat to those walls, and one of two things may happen. If it is dense and contains a lot of heat, it might melt the walls, and if it is rarefied and contains just a little heat, it will lose its heat to the walls, cool down and stop being a plasma. So how can you contain a plasma? Two ways have been tried —Dr Stern denied both.

I have given an answer to this question in the previous example of a marshmallow machine. Its shell is indispensable to the formation and existence of the sugar ball, **but the sugar ball and the shell do not need to be in contact; the temperature between them will not transfer, the role of the shell is only limited to change the direction of the particle motion.** Although, in this case, the sugar ball and the shell are of ordinary temperature. This can also be true of the stars if their shells are of a high temperature, which they are.

Take the Sun as an example. Its most primitive shell is the Oort cloud, and its inner edge is the Kuiper Belt, which keeps a distance of 30 AU from the Sun. Much of the Sun's radiant energy is intercepted by the corona and the recently discovered 'wall of fire' between the Oort Cloud and the Kuiper Belt, the Sun's two outer shells. The shell close to the Sun's surface is its photosphere (and chromosphere), which effectively blocks the outflow of the Sun's ions.

This could be controversial. Standard cosmology suggests that stars rise in temperature step by step from the outside in, with the core reaching its highest temperature of 20 million degrees Celsius, making them powerful radiators. However, the photosphere of the Sun is only 6,000K, which means it cannot act as a shell at all.

In the previous discussion about sunspots, I have reversed this view. The temperature of stars does not increase from the outside in, it decreases instead. The core temperature is the lowest, and nuclear fusion does not exist and cannot happen. **Thus, not only the 'wall of fire', the Oort cloud, the Kuiper Belt, the corona and the photosphere, but also every layer inside the Sun plays a role in heat preservation, and the Sun as well as the Solar System can keep its mass and energy steadily for billions of years.**

Its only loss is radiation light, which is formed by some high-speed ions in the photosphere escaping from the Sun, and the Sun depends on them for luminescence. Though this is an enormous loss in our eyes, it is insignificant compared with the Sun's total mass and can be ignored. In fact, most of them are blocked by two external hot shells, and are involved in hot shell construction, or simply return to the Sun.

A plasma is a semi-mass-semi-energy stuff. We might as well think of it as bare material energy block, or an ion resonator, or an electricity chunk wrapped by hot shells. Their changing law is opposite to that of atomic matter. The reason is very simple. **Ions need a wider moving space than atoms.** The looser the space, the higher their moving speed as well as temperature, and vice versa. Take the sun as an example, its gaseous plasma, namely the corona, has the highest temperature; Its liquid plasma, that is, its photosphere and surface layer, comes second in temperature; Its solid plasma, that is, the core of the sun, has the lowest temperature, and **the overall temperature of the sun drops from outside in, contrary to the density of its ions.**

ARGUMENT

"In the sun, for example, the temperature, which is only 6000°C at the surface, **increases gradually inward reaching in the center the tremendous value of 20 million degrees**. This figure can be calculated without much difficulty from the observed surface temperature of the solar body and from the known heatconducting properties of the gases from which it is formed. Similarly **we can calculate the temperature inside a hot potato without cutting it**, if we know how hot it is on the surface, and what the heat conductivity of its material is." (Gamow,1947, p.316)

According to the **wick effect**, I don't think that the surface temperature of a potato that just taken out of boiling water is lower than its internal temperature—Can it reach 100 degrees Celsius as its skin? Undoubtedly, the opposite is true. Note. The potato is an atomic substance, not the same sort of stuff as the sun, yet it has the wick effect.

Plasma ➤ White Dwarf ➤ Black Dwarf ➤Nuclear Bomb

Most people think that, except for supernovae, stars don't explode. They end up as collapsar, white dwarfs or black holes. CFVT offers different descriptions. If the Sun, for example, never exploded, its angular momentum will be preserved; it will continue to contract, the spin rate will accelerate, and the entity will change accordingly. Its gas and liquid layers are then compressed into a solid, forming a dense White Dwarf star, where the plasma still has room to move. But instead of losing its internal spin, it gets stronger and continues to contract until the plasma is pinned down and stops moving, turning the Sun into an icy black dwarf star. A black dwarf is a mass of stationary, non- luminous plasma.

Now let's imagine taking a cubic meter of plasma from the Sun's core, then compressing it into a tiny white dwarf, then a black dwarf, and then preserving it in a tightly sealed shell, which is a nuclear bomb. Through this, **we can think of a nuclear bomb as a miniature black dwarf.**

A nuclear bomb is the cosmic energy that is bound, and its explosion is the process of cosmic energy release, a process that is contradictory to the previous description. Under the action of high temperature, the shell of this miniature black dwarf will be destroyed, then it will be transformed into a tiny white dwarf, liquid plasma and ions ball, in turn. A small part then becomes a quark and they polymerize with environmental substances to produce new substances in which existing substances, including human bodies and artefact, are destroyed.

Such a bomb-making process is not realistic, but it is in common sense, better than popular belief. The popular theory is that a nuclear bomb explodes when neutrons are released from the nucleon by high temperatures, triggering chain reactions and releasing energy. What is this energy? There is no good answer for this yet.

This section proposes a new way for the formation of big

dwarfs and small black holes. Whether it is correct or not remains to be confirmed by future astronomical observations, which needs the support of observers who believe in CFVT.

A Nuclear Bomb— a Miniature Sun Ready to Explode

The invention of the nuclear bomb began with the study of atomic radiation. Scientists agree that atomic radiation is energy, not matter, and that atoms contain more powerful energy than any chemical energy. Unleashing it was the top task of 20th century scientists. But they immediately found the path of discovery circuitous.

"The truth is, however, that the effectiveness of charged nuclear projectiles produced in various accelerating machines is much smaller than that, and usually only one projectile out of several thousand can be counted upon to produce a nuclear crack-up in the bombarded material. The explanation of this extremely low efficiency of atomic bombardment lies in the fact that **atomic nuclei are surrounded by the envelopes of electrons that have the power to slow down the charged atomic projectiles moving through them.** Since the target area of the atomic envelope is much larger than the target area of the nucleus and since we cannot, of course, aim atomic projectiles directly at the nucleus, each such projectile must necessarily pierce many atomic envelopes before it will have the chance to deliver a direct blow to one of the nuclei. The situation is explained graphically in Figure where atomic nuclei are represented by solid black spheres and their electronic envelopes by lighter shadows. The ratio of atomic and nuclear diameters is about 10,000 so that the target areas stand in the ratio of 100,000,000 to 1. On the other hand, we know that a charged particle passing through an electronic envelope of an atom loses about one hundredth of one per cent of its energy, so that it will be stopped completely after passing through some 10,000 atomic bodies. It is easy to see from the above quoted numbers that only about 1 particle in

10,000 will have a chance to hit the nucleus before all its initial energy has been dissipated in the atomic envelopes. **Taking into account this low efficiency of charged projectiles in delivering a destructive blow to the nuclei of the target material we find that in order to transform completely 1 g of boron, we must keep it in the beam of a modern atom-smashing machine for the period of at least 20,000 years!"** (Gamow,1947, p.180-181)

"Everybody who reads newspapers knows that nuclear energy, or atomic energy as it is commonly- called, can be released through the fission process of uranium nuclei discovered by Hahn and Strassman late in 1938. But it would be a mistake to believe that the fission itself, that is, the splitting of a heavy nucleus into two nearly equal parts, could contribute to the progressive nuclear reaction. In fact, the two nuclear fragments resulting in fission carry heavy electric charges (about a half charge of the uranium nucleus each**), which prevent them from approaching too close to other nuclei. Thus, rapidly losing their initially high energy to the electronic envelopes of neighboring atoms, these fragments will rapidly come to rest without producing any further fissions.**

What makes the fission process so important for the development of a self-sustaining nuclear reaction is the discovery that before being finally slowed down each fission fragment emits a neutron. This peculiar aftereffect of fission is due to the fact that, like the two pieces of a broken spring, the two broken halves of a heavy nucleus begin their existence in a state of rather violent vibration. These vibrations, which are not able to cause a secondary nuclear fission (of each of the fragments into two), are, however, strong enough to cause the ejection of some nuclear structural units." (Ibid., p.182-183)

"It can be concluded from the general theory of nuclear structure that the fission effectiveness of neutrons increases with the increasing atomic weight of the element in question, coming fairly close to a hundred per cent for the elements near the end of the periodic system." (Ibid., p.184)

Gamow Let's know that bombarding a nucleus with charged particles to release the energy it enshrouds would be repulsive, inefficient and impractical. Using a neutron chain reaction, and using heavy atoms as material, the bomb was finally built. Heavy atoms have fission rates of up to a hundred percent.

But Asimov believed that the efficiency of atomic bombs and hydrogen bombs is still very low, with nine hundred and ninety-five over one thousand of nuclear material not working.

"Meanwhile the fission bomb had been reduced to triviality. Human beings had succeeded in setting off another energetic nuclear reaction which made much more devastating bombs possible.

In the fission of uranium, 0.1 percent of the mass of the uranium atom is converted to energy. But in the fusion of hydrogen atoms to form helium, fully 0.5 percent of their mass is converted to energy." (Asimov, 1993)

Here I think it is necessary to further explore the mode of nuclear energy and its release.

"the term 'Nucleonics' should be understood to apply to the science of practical applications of nuclear energy liberated on a large scale. We have seen in the previous sections that the nuclei of various chemical elements (except silver) are overloaded with tremendous amounts of internal energy that can be liberated by the processes o£ nuclear fusion in the case of lighter elements, and by nuclear fission in the case of heavier ones. **We have also seen that the method of nuclear bombardment by artificially accelerated charged particles, though of great importance for the theoretical study of various nuclear transformations, cannot be counted upon for practical use because of its extremely low efficiency.**" (Gamow,1947, p.181)

This notion is based on Einstein's theory and formula of mass-energy conversion, and its calculation unit is electron volt, erg, ton·TNT, etc. However, up to now, no one is sure what nuclear energy is.

Specifically, energy and field are inseparable, and the energy theory established from particle theory cannot be perfect, and the existing fission theory and fusion theory are imperfect.

"It must be remembered, however, that neither the fusion of two light nuclei, nor the fission of a heavy nucleus would normally take place unless we did something about it. In fact, to cause the fusion of two light nuclei we have to bring them close together against the repulsive forces interacting between their charges, and **in order to force a heavy nucleus to go through the process of fission we must start it vibrating with a sufficiently large amplitude, by giving to it a strong tap.**" (Ibid., p.168)

According to CFT, nuclear energy is light trapped in a nucleon, a tiny, intense field surrounded by a unified field of electrons that together make up atoms. We know from the works of Gamow and Asimov and other scientists that high temperatures and strong magnetic fields cause nucleons and electrons to separate and become ions. The sun is a plasma; **A nuclear bomb, both atomic and hydrogen, is also a plasma, a small artificial sun when it explodes. The most obvious difference between the two is that nuclear bombs explode immediately after they are launched, whereas the present sun does not, due to its shells.**

An explosion is a sharp rise in the temperature inside the object, a sudden increase in tension, breaking through the constraints of the shell and causing the object to split. Notice that this book describes temperature as particle velocity.

According to CFT, the sun and other stars are relatively

cool plasma inside and relatively hot plasma outside, with material spewing out from the surface and a core that does not explode. The corona is its close-in outer shell, effectively blocking its mass and energy loss.

A nuclear bomb. Its core is a pure, dense nucleus ball wrapped in an invisible electric field. Once triggered, first, the high temperature of the outer explosive dissipates the electric field and turns the core into a plasma. Second, with no shell to restrain it, the plasma disperses into higher temperature torches, and its heat disintegrates most of the ions, freeing quarks to fly away at the speed of light. In the third, plasma torches & quarks interact with air and ground objects to form fireballs, shock waves and fatal radiations.

We can also make a simpler description of the explosion of nuclear bombs. Each nuclear bomb experiences two explosions. The popular idea is that the first explosion creates high temperature, activates chain reaction or nuclear fusion, and produces a second, stronger explosion.

CFT described this process as: **the first explosion completely eliminated the electric field of the nuclear bomb, and all atoms were converted into ions and photons, which caused the resonance of environmental substances, that is, the second explosion.**

Both atomic and hydrogen bombs go through the same process, but with different efficiency. The reason is that atomic bombs are made of heavy atoms such as uranium or plutonium, which drain part of their energy to convert them into hydrogen ions. Hydrogen bombs are made from hydrogen isotopes. Hydrogen, the simplest element, can be converted directly into hydrogen ions at high temperatures.

NOTE. In 'The Atomic Bomb Is Also a Fusion Weapon', I follow the corpuscularianism to explain the nuclear bomb. In this paper, corpuscularianism and fission and fusion theory based on it are completely abandoned and CFT is adopted, which is a more eloquent answer.

A proposition of CFT, 'the solar plasma emits light by itself and does not need nuclear fission or nuclear fusion.'

Nuclear bombs work the same way. The sun is a giant ion ball with no electron field, if it were suddenly removed from its shells, viz., its photosphere, chromosphere and corona, it would instantly turn into a giant hydrogen bomb and explode!

Genesis Conditions

The previously stipulated definition of matter is also the way of energy-mass conversion. **How energy is converted into matter has long been a puzzle, but the answer is as simple as this: Don't allow energy (quarks) fly freely, wrap it up. Even if it's a hypothesis, you can't find a better one.** If you believe it, you have cracked the biggest codes in the physical world. What's even more amazing about this definition is that it can also provide the answer to the creation of everything, in both the micro and macro ways, starting with atoms.

The core of an atom is a proton, the proton is matter, and its interior is energy (quarks). Nuclear fusion produces not only energy but also new matter. Hydrogen is the simplest atom, and nuclear fusion allows four individual hydrogen atoms to break out of their confinement and instantly tie themselves back together; the four tiny energy fields coalesce into a stronger field, the helium atom. Other atoms aggregate in a similar way. That is the micro way, which is actually the way basic elements are generated.

The macro approach targets stars. Stars are matter. Although, I regard them as big nucleon now. It consists of plasma, which itself is semi-matter, similar to quarks in a nucleon. The space within a nucleon is a field and static, but the inner space of stars is vast—plasma is imprisoned relatively loosely. Although crowded, there is still room for ions to move fast, so they cannot coalesce into atoms, but the aggregation movement can occur on their ejecta.

Due to relatively little mass & relatively low temperature, small stars cannot continue to emit light but polymerization can be effective. This is a process in which ions lose their freedom and energy is gradually trapped. It triggers a phase transition, producing more and more liquid and solid bodies, and the phase transition becomes faster and faster, and more and more intense. This is how we now see the liquid hydrogen layer and diamond sea on the Jovian stars. However, the contraction movement will raise their temperature and return them to the ionic and plasma state again. Earth-like planets do not have this kind of cycle, and the phase change can proceed steadily. They are planets that successfully trap most ions permanently.

Unable to get rid of the high temperatures, the Jovian stars are still ionic in nature, and the atomic polymerization, though occurring, cannot continue. Earth-like planets are essentially atomic planets, where the splitting and merging of atoms can continue to take place, causing the emergence of more kinds of atoms, molecules and objects with more complex structures, until eventually intelligent creatures and humans are born.

From this, we found an astrophysics law: **high temperature achieves unity, low temperature produces diversity** — this is also the guiding ideology of the GUT (Grand Unified Theory).

G. Gamow wrote in his book: "physicists to a striking conclusion that: the common chemical elements that form the body of the Earth constitute only about 1 per cent of the solar mass, the rest being almost evenly divided between hydrogen and helium with a slight preponderance of the former. Apparently this analysis also fits the constitution of the other stars." (p.305) This fact supports the above law.

I have long applied the concept of evolution in biology to cosmology, especially the evolution process from nebula to biology, which can now be further expressed as the process of plasma movement and change, which only happens on planets. Its reason is simple and beyond doubt: **Ions are the products of atoms splitting at high temperature, and the indispensable condition for ions reduction to atoms is low temperature.**

Conditions for this are not yet sufficient for the Jovian planet, and the reduction can only occur during its dark period, and will be interrupted when it turns to the luminous period. It is only on the Earth-like planets can this movement last and the evolution process be complete, from low to high, from simple to complex, and the Earth is the perfect model.

Now we observe that **plasma is the ultimate source of liquid hydrogen, diamonds and other metals on the Jovian planets, and lava, oceans, land, minerals, plants and animals on the Earth-like planets. From plasma to human, it is the most complete and perfect universe material evolution movement.**

REFERENCE
The Evolution formula of the 19th Century British Thinker Herbert Spencer.

"So Spencer offers us his famous formula of evolution, which made the intellect of Europe gasp for breath, and required ten volumes and forty years for its explanation. Evolution is an integration of matter and a concomitant dissipation of motion ; during which **the matter passes from an indefinite, incoherent homogeneity to a definite, coherent heterogeneity**; and during which the retained motion undergoes a-parallel-transformation...the nebula is shapeless, nebulous ; and yet out of it come the elliptical regularity of the planets, the sharp lines of mountain-chains, the specific form and character of organisms and organs, the division of labor and specialization of function in physiological and political structures, etc. And the parts of this integrating whole become not merely definite but diverse, heterogeneous in nature and operation. The primeval nebula is homogeneous...these are the foci of the orbit of evolution. Whatever passes from diffusion to integration and unity, and from a homogeneous simplicity to a differentiated complexity (cf. America, 1600-1900), is in the flow of evolution; whatever is returning from integration to diffusion, and from complexity to simplicity (cf. Europe 200-600 a. d.), is caught in the ebb of dissolution." (Durant,1991, p.341-343)

The Way of Mass-to-Energy Conversion

This is a process of destroying the shells of matter in turn, from the shell of an object, to the shell of a molecule, of an atom, of a nucleon, of a proton. Accordingly, the movement speed gradually accelerates.

Again, take water as an example. Ice is an object wherein the movement of water molecules is suspended, leaving only slight vibrations. After melting the ice, the outer shell of the frozen water is eliminated, and its molecules can flow freely. Then, when water is splashed on fire, the molecular shell of water is eliminated, and water atoms can fly freely. With further heating, the atomic shell of water is eliminated, and hydrogen ions and oxygen ions can fly faster. When heated again, the quarks are freed from the ions and fly away as light. In this series of changes, the motion speed changes from nanometre-speed to light-speed. All along, we don't see the energy, we just see the particles speeding up, and their velocity is energy.

In contrast to the energy-mass conversion process, mass-energy conversion is a decomposition process in which matter is divided into parts and finally disappears — it is all converted into energy, which is light. Each step of this process requires external energy. The principle of 'decomposition of energy consumption' is confirmed here.

The premise is to confirm the cosmic force. **The quark is a cosmic force; it is also a kind of photon. After destroying the layers of material that bind it, the conversion of mass to energy naturally occurs — free quarks fly forever and never rest.**

This is the reason why popular ideas are confusing. It is impossible for a ruler, a yardstick, a spacecraft, a person, etc., to fly as a whole and realize mass-energy conversion by increasing speed.

All objects are composed of atoms. Their particles are

imprisoned and cannot be freely exercised. Their flight requires external energy for power. To make any object move at the speed of light, an infinite amount of external energy is required. Therefore, no object can reach the speed of light, let alone exceed the speed of light. Time travel, returning to the past or entering a future world, can only remain an unrealized scientific fantasy forever.

Atoms, Corpuscles, and Reductionism

The real whole might well be, we conceive, an indivisible continuity. The systems we cut out within it would, properly speaking, not then be parts at all; they would be partial views of the whole. And, with these partial views put end to end, you will not make even a beginning of the reconstruction of the whole, any more than, by multiplying photographs of an object in a thousand different aspects, you will reproduce the object itself. So of life and of the physico-chemical phenomena to which you endeavor to reduce it.
(H. Bergson, *Creative Evolution*)

"So far, the unification of science has been seen in the reduction of all sciences to physics, the final resolution of all phenomena into physical events. From our point of view, unity of science gains a more realistic aspect. A unitary conception of the world may be based, not upon the possibly futile and certainly farfetched hope finally to reduce all levels of reality to the level of physics, but rather on the isomorphy of laws in different fields...We come, then, to a conception which in contrast to reductionism, we may call perspectivism. **We cannot reduce the biological, behavioral, and social levels to the lowest level, that of the constructs and laws of physics.** We can, however, find constructs and possibly laws within the individual levels."
(Bertalanffy, 1969, p48-49)

The history of reductionism can be traced back to Democritus' atomism. In his opinion, the atom is the smallest unit of matter and cannot be divided; everything is made up of atoms. Atoms are discrete, so atomism becomes the basis of corpuscularianism. And, as we have established, its opposite is holism.

The Greeks made many important contributions to

mathematics and influenced philosophy and science. Corpuscularianism is naturally combined with mathematics, which has become its handy tool. The two promote each other and develop together, expanding into the mainstream theory of Western science, collectively called reductionism. Whereas holism is not convenient for mathematical calculation and has gradually been abandoned.

Newton was an advocate of the theory of particles, his corpuscular theory of light was obviously corpuscularianism. In Europe, the application scope of corpuscular theory soon went beyond matter and was expanded to include nonphysical entities such as light, heat, and force. The theory of phlogiston, caloric theory, photon theory, graviton theory, quantum theory, and string theory were produced. Newton's argument that 'particles may be divided and actually separated to infinity' (The Principia, 389), is one basis for them. They were considered different types of corpuscularianism.

However, reductionists are not satisfied with this. They also give corpuscles free will. Typical examples are Ernst Haeckel's cell soul theory, Francis Crick's atom consciousness theory, Richard Dawkins's 'selfish gene' theory, and quantum scientists' 'quantum consciousness' and 'quantum soul' theory. The Big Bang cosmology is the latest and most advanced version of it. The universe has finally been reduced to a particle, or a singularity.

Western science consists of almost all reductionist works, and discreteness has become its main feature. Some theories that need a holistic view and continuity, such as energy theory, field theory, and cosmology, are therefore difficult to develop or are being led down a crooked road. This is why scientists ignore the swirling phenomena and light field that fill the sky, and why they reject the vortex theory.

Generally speaking, reductionism (induction) is a bottom-up way, while holism (deduction) is a top-down way. The purpose of scientific research is to discover the truth, and the latter is far superior to the former.

However, reductionism is an indispensable tool in scientific research, and it is used in this book also. This discussion is intended to introduce holism, improve reductionism, and combine the two. Here, we have to go back to Plato.

Plato's theory of ideas is a typical holism. Holism looks at things in order from big to small, from the ideal thing to the defect thing and through the former to the latter, to ensure the right direction. Plato, taking justice as an example, pointed out that there is justice of one man and there is justice of a whole city too. "So then, perhaps there would be more justice in the bigger and it would be easier to observe closely. If you want, first we'll investigate what justice is like in the cities. Then, we'll also go on to consider it in an individual's, considering the likeness of the bigger in the idea of the littler?" (*The Republic*, 369a)

This is easy to understand, but modern scientists do the opposite; they take Aristotle's approach. They focus on the small, ignore the big, or calibrate the big with the small, and build the world's largest machine to observe the tiniest things in the universe, corpuscles or their shadows of all kinds. It may be time to rethink this approach.

Experimental philosophy is inseparable from experimental tools. Modern science has become increasingly dependent on big machines, and scientists have become artisans, as Einstein put it, far removed from the path of wisdom.

"Some Greeks, notably Plato, had other reasons, equally weighty to them, for imposing the restriction. The introduction of more complicated instruments which might be adequate to the solution of the construction problems called for manual skill unworthy, in their opinion, of a thinker. Plato said, further, that by using complicated instruments 'the good of geometry is set aside and destroyed, for we again reduce it to the world of sense, instead of elevating and imbuing it with the eternal and incorporeal images of thought, even as it is employed by God, for which reason He always is God.'" (Kline, 1953, p50-51)

Utilization and Conditions of Material Energy

All burning and luminous movements are mostly particle movements, or vibrations, characterized by high temperatures and high speeds, which generate or consume cosmic energy. Our daily activities, such as walking, labour, physical exercise, driving vehicles, etc., and nature's movements, such as wind, rain, tides, earthquakes, etc., are mostly molecular and object movements, characterized by low speeds and no light, which generates and consumes material energy.

What we often use is material energy. Even if we use cosmic energy, we often need to convert it into material energy first. This involves the research of the utilization and conditions of material energy.

We need to be clear here, what do we mean by the use of energy, both cosmic energy and material energy? Take the old coal-fired power generation as an example. How does the energy in coal turn into electricity?

From the perspective of corpuscular motion, the process involves using artificially high temperature to catalyse coal combustion. That is, input some ions moving & vibrating at high-speed into the coal pile, let them break through the molecular shells and atomic shells of coal, become ions, which push the chain reaction, so that more ions in the whole pile of coal are gradually released. Most of these ions radiate away as light (returning to the light field), or as elements combine with air molecules. The remaining ions push other particles to move or interact with them in various forms, including pushing the atoms of coal (that have not been broken) to collide directly, or pushing them to combine with oxygen, and collide with the boiler wall on a larger scale, pushing the metal atoms at the bottom of the boiler wall to move & to vibrate at high-speed. Then there is a series of movements: high-speed movement of water molecules in the boiler → high-speed movement of steam in the pipeline → steam engine → generator, which finally generates a electric current.

In this process, we can see that the carriers of energy are gradually consumed, and a series of particles and particle objects (steam engine, generator) movements are accelerated, but **never see energy.** It is like a powerful invisible man, and we feel his existence intensively, but we can't see his body in any way, even though he is propelling this series of motions from beginning to end.

Other energy utilization are much the same, we can't *see* any energy —All energy are invisible light.

From the above discussion, we summarize the following conditions for energy utilization:

First, **a high degree of purity.** An object is composed of molecules and atoms. In physics, the release of material energy is mainly manifested as the acceleration of the motion of molecules and atoms. Relatively pure substances, such as natural gas and gasoline, that have molecules and atoms which move vibrate at the same speed because they are relatively pure, can give full play to their functions by resonance. While impure substances, such as rocks and soil, that have molecules and atoms which have very different velocities, cannot move in unison and will interfere with and offset each other, thus making the use of material energy difficult or even impossible.

Second, **wider space.** The accelerated motion vibration of molecules and atoms requires a wider space, which causes the volume of the object to expand. If space is restricted by the natural environment or artificially suppressed, the release of material energy is restricted. That is the reason why fuel containers, such as gasoline tanks and gasholders, are safer when they are fuller.

Third, **polymers**. The release of material energy is a splitting movement for carriers and a polymerization movement for other objects. The most common is polymerization with oxygen in the atmosphere, to generate new matter. Covering open flames with oxygen- free gas or liquid is one of the firefighting methods commonly used by the fire brigade. The reason for doing this is to limit the polymerization of substances with oxygen and also limit oxygen's motion space.

Fourth, **controlled facilities.** The purpose of using material energy is to release the trapped particles, reducing them to cosmic energy. Uncontrolled, rapid release of energy is often a disaster, the bomb explosion is an example. Only when energy is under control, can it be safely to use.

Although the universe is rich in energy, we can only use the material energy on Earth and subject it to the foregoing four conditions, which is why the energy crisis soon followed the arrival of the industrial era.

NOTE

1. in the view adopted by this book, the Earth is an outcrop of the Sun, and all matter and material energy on the planet are the products of the plasma phase transition, including air, water, rock, soil and minerals, from which organic stuffs are later products. The main material energies, such as coal, oil, natural gas and so on, are also the conversion of plasma and the components of the original crust; they, in their own way, participate in and affect the formation of plants and animals, which cannot be the source of material energies. As a result, energy reserves are much higher than previously thought and the energy crisis is expected to ease.

2. The human body burnt by high voltage electricity has similar symptoms and similar principles to those of spontaneous combustion. I interpret the two events as the loss or destruction of the atomic shell of the human body; the sudden release of the energy trapped in the atoms of the human body, **whose partial or whole changes into plasma, and interact with the air, resulting in burning. It needs to be understood that plasma is the original material of everything, including our body.**

Spontaneous Human Combustion (SHC) — Ions & Quarks Revel in Freedom

***Electricity I am.** Or, if you wish, I am the electricity in the human form. **You are Electricity too;** Mr. Smith, but you do not realize it... **Man's body and brain are made from a large amount energy;** in me there is the majority of electricity. The energy that is different in everyone is what makes the human 'I' or 'soul'. For other creatures to their essence, 'soul' of the plant is the 'soul' of minerals and animals.*

Nikola Tesla (1899, interview with journalist J. Smith)

Like everything else, the human body is the arsenal of atoms, storing up vast quantities of energy. Once the compartments inside the arsenal are broken, and the guardians of the atom, the electrons, withdraw and become active ions, a horrible event will happened. Atomic bomb explosion and Spontaneous Human Combustion are two typical cases.

SHC phenomenon is rare, but the cause of its occurrence remains a mystery, arouse wide concern. Here is a typical description on it.

SPONTANEOUS HUMAN COMBUSTION | PARANORMAL MYSTERIES

By Charlene Lowe Kemp / April 27, 2021

For 300 years there have reportedly been over 200 cited cases of a mysterious, rare, phenomenon that kills people — one in which the body is said to 'ignite' by itself and burn its victims alive, yet mostly leaves their surroundings alone. What is the cause of Spontaneous HumanCombustion?

SHC is a phenomenon that has baffled forensic experts, scientists and paranormal investigators for years. It is said that the fire originates from within the human body. What makes this even more bizarre are two things: **usually only** the body and clothing are found burnt, with nearby furniture and items largely untouched and usually damage is found on the ceilings above the cremated remains.

Also bizarre is the fact that usually the victim is alone and appears to not have struggled.

In many cases, only a limb is found intact and uncharred—most times a leg or foot. In some cases the skull has also shrunk.

HERE IS TWO TRUE ACCOUNTS OF SHC... (https://paranormalhauntings.blog/2021/04/27/)

Some people attribute the phenomenon to alcohol, methane, smoking, etc. Others attribute it to body fat and clothing, suggesting the wick effect. I appreciate with Larry E Arnold, director of ParaScience International and prominent researcher of fire phenomena, he proposes a theory in his 1995 book *Ablaze!* for the idiopathic thermogenic event involving a biologically-induced nuclear explosion. But it is not as good as my **atom-ion reductionism**, which is a theory I recently conceived.

This theory comes from my idea of how a tungsten filament electric lamp generates light. I think the reason of an electric bulb glow is that when the circuit is switched on, the current forces all the filament electrons disengage from atoms, to join the system electric field, and all filament atoms become plasma like the sun. **Atoms are stable, vibrate slightly, and do not emit light; Plasma is a mixture of energy and matter. Its vibration is a thousand times stronger than that of the atom, gives off heat & light.** Ions impinge also produce large numbers of free quarks (photons), which form the main source of light. The intenser the current, the more free quarks and the brighter the lamp. When the circuit is disconnected and the lights go off, the plasma instantly replenishes electrons from the wires and returns to tungsten atoms. This is an atom-ion-atom reduction cycle, or A-I cycle for short. **It is ionization, not combustion, because there is no oxygen in the bulb.** It also proves that the luminescence of plasma, such as the sun & the lava, has nothing to do with thermonuclear reaction or nuclear fusion.

In addition to tungsten, metals such as platinum and plant fibers can act as filaments, Thomas Edison had used more than 600 different stuffs in his experiments. But most of them could not withstand the shock of current and would soon fuse. The fluorescent lamps and neon lamps, emit light without filaments. They are infused with gases such as nitrogen, helium, neon, argon, krypton, and xenon. From the Internet I found some people don't apply current, but simply place a coil between two magnets, which can light an incandescent bulb. (It has not been accepted by the scientific community) All of these luminescence phenomena work on the same principle, viz., ionization & resonance.

SHC has one thing in common with electric lamp, that is, part or whole of the protons of human body lose their electronic shells and become ions, and some quarks are free, the human body will instantly be turned into a plasma, which will scorch the body and never have the chance of reduction, namely, it has no A-I cycle. Part of quarks and ions that escape the body interact with the air to form flames, leaving behind oil stains, ashes and wreckage. That's why SHC couldn't last, and the victims didn't struggle— they had no time to struggle before they were ionized.

Quark is a kind of photon, which is cosmic energy. Any object, once the electrons are lost and the electric field is weakened or removed, its quarks will rush out instantly and the object will be annihilated. It's a fastest and most thorough motion than any burning. The reason why we can see the flames and wreckage of SHC is that its electric field has not completely removed, and most ions & quarks have not freed, or have not had time to free themselves from atoms, human body is destroyed or partially preserved, mostly are hands or feet, by a small amount of ions & quarks that freed earlier. This is just like the early atomic bomb explosion, in which the chain reaction was interrupted by a small amount of energy just released, leaving most of the atoms in flames before they could release their internal energy.

As the previous discussion shows, not only electricity but also magnetism can drain electrons from the material or gas inside the bulb, turning atoms into ions & photons. (Reference articles: The Simplest and Clearest Principle of Electromagnetism) **Why does the human body lose electrons and cause fire without contact with any sources of fire, power or strong magnets? The answer is, too many positive ions (cations).** This is a very complicated process.

We know, water makes up the largest percentage of body volume it is not pure water but a mixture of water, cells, proteins, glucose, lipoproteins, electrolytes and so on, generally referred to as body fluids, the main sources are air, drinks and food.

Many normal substances exist in the body as ions. Common examples include sodium, potassium, calcium, chloride, and bicarbonate. These substances are known as **electrolytes,** which plays a role in conducting nervous impulses, contracting muscles, keeping our hydrated and regulating our body's PH levels, etc.

Electrolytes are atom and molecule that have gained or lost one or more valence electrons, giving them net negative or positive charge, called anions and cations respectively. Because they have opposite electrical charges, cations and anions are attracted to each other. Cations repel other cations and anions repel other anions.

Due to congenital or acquired eating habits and other reasons, that makes the amount of cations in the body too high. When reaching a critical point, they will tempt the electric field of the body to stand alone, not as a shell for the nucleus, but as its opposing force. At this point the human body becomes plasma and SHC occurs.

My idea can be confirmed from the events investigated, and can be used as explanations for these events in turn. For example, why do most SHC events occur indoors when the victims sleep or sit and rest?

Scientists tell us that the human body is often dominated by cations, which has reductive property, that is, robbing electrons from normal atoms, and needs to be replenished anions from the breath to eliminate cations reductive impulse. Anions in outdoor atmosphere is several times to dozens of times higher than that in indoor atmosphere. Bedroom are often small in space, and the air does not flow, lack anions especially. Public places are wider, but if there are too many people or the air is polluted, it will also lack anions. Therefore, one staying indoors and crowded public places, such as dancing parties, for a long time, is prone to anions deficiency.

In addition. When people are at rest, the cells membranes gradually open, and K+ on the inner surface of the cell membrane will go out, making the total amount of cations in body fluid greatly increase.

I would like to make a supplement here. The cation has an ally, the **free radical**, which is an atom, molecule, or ion that has an unpaired valence electron, it is also keen to plunder electrons. They follow the blood throughout the body, performing their functions while constantly attacking the body's normal atoms and molecules, especially blood vessels and bones. "There are many types of radicals, but those of most concern in biological systems are derived from oxygen, and known collectively as reactive oxygen species. Oxygen has two unpaired electrons in separate orbitals in its outer shell. This electronic structure makes oxygen especially susceptible to radical formation." (http://www.vivo.colostate.edu) **When antioxidant is too weak in a person's body, radical cations overwhelm radical anions, the probability of SHC will go up.**

Related to this is **electronegativity,** which is "a chemical property that measures how likely an atom is to attract a shared pair of electrons towards itself in a covalent bond." ('What Is Electronegativity? Definition, Chart, and Trends'. Posted by Hayley Milliman. Sep 23, 2019, GENERAL EDUCATION) We need oxygen all the time. **Oxygen is essential for life. But oxygen is extremely electronegative,** that is, its ability to grab electrons is so strong that it ranks second on the periodic table. This means that high levels of oxygen in the body increase the risk of SHC.

Some people are born with excess cations and develop SHC shortly after birth or when they are young. Others, for a variety of reasons, have large amounts of cations build up in the body, culminating in the occurrence of SHC in the situations mentioned above.

The SHC incidents are notable, the events listed on the web are reliable, and there is a growing trend, statistically,

related to the decline in our food and air quality. However, so far such kind of incidence is few, has not caused much impact on human survival, scientists have not listed it as a subject of scientific research, but our discussion is still meaningful.

In the modern world, the admonitions we often hear are, 'Aerobics' & 'Get out in nature and soak up more anions'. This discussion has strengthened our confidence in this creeds. More importantly, it deepens our understanding on the cosmic energy.

ARGUMENT

One consensus among atomic scientists is that protons are unbreakable, that quarks cannot be released easily from matter. Gamow believed that the nucleus is also unbreakable—It should be easier than breaking protons.

"In order to attain complete thermal dissociation of matter, that is to break up the nuclei themselves into the separate nucleons (protons and neutrons) the temperature must go up to at least several billion degrees. Even inside the hottest stars we do not find such high temperatures, though it seems very likely that temperatures of that magnitude did exist several billion years ago when our universe was still young." (Gamow,1947, p.198-199)

But what could sunlight, nuclear blast light, lightning light, candlelight and SHC light be but quarks released from matter? Scientists admit that matter has enormous internal energy, but no one has given it a name. This book, for the first time, explicitly stated that it was quarks, viz., light trapped in the nucleus of an atom.

The above facts show that nuclear energy does not need extremely high temperature and great extra energy to be released. We release it and use it every day.

Dark Matter Is Mainly Composed of Nebulae, Small Stars, and Black Holes

We see matter and we miss energy; we think that we know what matter is; but when at the heart of the atom we find energy, we are bewildered, and our categories melt away.
(Durant, 1991, p.494)

Dark Matter, Rodney A. Brooks wrote,
"As early as 1933, astronomical observations showed that there is more matter in the universe than was previously thought. In fact, astronomers now believe that ordinary matter constitutes only about 5% of the total mass of the universe, while something called dark matter makes up 27% of the total mass. (The rest is made of a related substance called dark energy, which we won't go into.) This conclusion is based on its apparent gravitational influence on other objects. **Dark matter is found primarily around galaxies and is believed to be a million times less dense than normal matter.** It is not 'seen' because it doesn't emit or absorb light — that is, it doesn't interact (or interacts very feebly) with the EM field.

QFT offers two possible explanations for dark matter. One possibility is a field suggested by Steven Weinberg and Frank Wilczek, with quanta called axions...The other possibility is a field suggested by an approach to QFT called supersymmetry. Either of these hypothesized fields would be sufficiently stable and interact feebly enough with normal matter to qualify." (Brooks, 2016, p.180-181)

In my opinion, **dark matter are celestial bodies without light, they do exist and is more abundant than matter.** Although most of them have not been observed, we are not difficult to deduce it.

The first source is nebulae. Astronomers have confirmed that nebulae are the source of stars. According to the phenomenon that the origin is always more than its products, I believe **the total number of nebulae in space is far greater than the total number of stars, but nebulae do not emit light and most of them are not observed.**

Scientists have long thought of The Oort cloud to be a remnant of the original protoplanetary disc that formed around the Sun approximately 4.6 billion years ago, with a thin space of matter. The latest Voyager 2 data, however, tell us that material outside the 'fire wall' (18.5 billion kilometres from the Sun) is 0.12 electron per cubic centimetre, and inside (17.9 billion kilometres from the sun) is 0.039 electron per cubic meter; this is nearly a three-fold increase in density, which is new evidence.

The second source is **small stars, Jovian stars;** as has been established, they are a kind of intermittently glowing star. Jovian stars have a much longer period of dimming than they do of luminescence and are much more numerous than the stars that continue to glow, making them another important component of dark matter.

Black holes are the third source of dark matter, and also of dark energy. **According to the new black hole theory, they are also evolved from nebulae and stars. They are numerous and cannot be seen until they ignite and glow.**

In addition, there are Earth-like planets and satellites. They can reflect starlight, which is originally visible, but because of their small size and long distance, they are not easy to observe and become dark matter.

Taken together, it is possible to say that dark matter exceeds matter. However, the object and function of dark matter here are completely different from popular ideas.

The previous discussion has confirmed that matter is atom, which is imprisoned energy, and free particles are not matter, but energy. Therefore, **even if such particles — gravitons, axions and sterile neutrinos — exist, they cannot be called matter or dark matter.** Only when they lose their ability to move freely and are imprisoned, that is when they form atoms and objects and are not observed by us for various reasons, can they be called dark matter.

In the Simplest Sense, Dark Energy Is Light and Black Holes

"Dark energy is a mysterious force making the universe fly apart faster and faster...**Dark energy is the name for whatever is causing this accelerating expansion. It dominates the cosmos, making up by the latest reckoning some 68 per cent of everything there is.** But what is it? There, physicists are stumped, even more so than with dark matter, that other mysterious cosmic apparition. At least we know dark matter gravitates. The effect of dark energy seems to be to oppose gravity, whose pull would otherwise tend to make the universe contract — so it doesn't seem to correspond to any physical phenomenon we've yet encountered. One possibility is that **dark energy is a vacuum energy of the sort that quantum particles might create by popping in and out of free space.**"

(Richard Webb. Dark Energy. https://www.newscientist.com/definition/)

A popular view on the internet, in terms of physical cosmology and astronomy, is that dark energy is an unknown form of energy which is hypothesized to permeate all of space, tending to accelerate the expansion of the universe. Cosmologists estimate that the acceleration began roughly 5 billion years ago. Before that, it is thought that the expansion was decelerating, due to the attractive influence of dark matter and baryons. The density of dark matter in an expanding universe decreases more quickly than dark energy, and eventually the dark energy dominates. Its two leading models are a cosmological constant and quintessence. Both models include the common characteristic that dark energy must have negative pressure; some cosmologists have labeled it 'gravitational repulsion', 'vacuum energy', or 'zero-point energy'. Negative pressure is thought to have no influence over the gravitational interaction between masses —which remains attractive — but rather alters the overall evolution of the cosmos at the cosmological scale, typically resulting in the accelerating expansion of the universe, despite the attraction among the masses present in the cosmos. The acceleration is simply a function of dark energy density. Dark energy is persistent; its density remains constant, i.e. it does not get diluted when space expands.

Assuming that the standard model of cosmology is correct, the best current measurements indicate that dark energy contributes 68.3% of the total energy in the present-day observable universe. The mass–energy of dark matter and ordinary (baryonic) matter contribute 26.8% and 4.9%, respectively, and other components, such as neutrinos and photons, contribute a very small amount.

From the foregoing, it is clear that cosmologists view dark energy (and energy) as an independent, massless force that pushes everything around and causes the universe to expand.

From the proportion of dark energy, it should be the CFT said cosmic energy, which exists in the form of a field, and is constantly diverging in the form of waves. It is not only invisible, but also endless. That is to say, it needs a rich and endless material source to sustain its existence. What is the source of dark energy?

CFC view is very clear: **the cosmic energy is light, it is composed of the light field, and from the radiation of all luminous objects, energy and matter cannot be separated.** The light field interacts with matter in the form of photons, that is, the photons velocity is transmitted to all the objects interacting with each other, or they combine with the object to form new substances. **All energy is invisible, and so is dark energy. All we can see is the fact that a bunch of things are moving & vibrating faster by the invisible energy.**

CFC argues that if material energy, including the energy trapped in atoms and plasma, is divided into visible energy, the energy outside of it is divided into invisible energy, i.e.

dark energy — mainly light and black holes. The popular energy-mass ratio will be reversed with material energy at 75% or more, and dark energy at 25% or less. The main influencing factor is the black hole.

Though cosmic energy — which is light — spreads over vast areas of space, its contribution to dark energy is limited. The Sun, for example, loses 4.2 million tons of mass per second, according to Asimov's calculations in *Asimov's New Guide to Science* (1993), and only about 1/33,000 of its total mass is lost over the past five billion years, because stars don't need thermonuclear reactions to shine. By analogy, for

13.7 billion years the amount of energy that the universe has received from the radiation of celestial bodies is the equivalent of, at most, 1/10,000 of their total mass, plus the energy from black holes, which accounts for more than 20%.

It is only when the energy trapped in matter and plasma are taken into account that the percentage of energy is 96% or even 99%. However, this material energy cannot be called dark energy. It is stationary and participates in a swirling motion in the form of atoms and plasma. Therefore, its polymerization force is not the driving force of the expansion of the cosmos, but the force of the contraction of the cosmos.

Anyway, since cosmic energy fills the whole space, we need to pay close attention to it. It is the background field, the light field, and acts primarily by energizing everything by increasing the speed at which molecules and atoms move, or by combining with matter to form new matter. Ordinary objects interact with light fields inconspicuously, and with large and hot objects, conspicuously. The Sun is a good example. The surface temperature of the Sun is about 6,000 degrees Celsius, while the temperature of the corona, which is located in the outermost layer of the Sun's atmosphere, reaches 1 million degrees Celsius. This is a phenomenon that has puzzled astronomers so far. The most plausible explanation is the enhancement effect of the light field that is proposed in this book. Not only the Sun's corona, but also the ionosphere above the Earth and the hot shells of all celestial bodies, can be regarded as the common product of the light field's and celestial bodies' radiation. This is also the effect that a pair of particles in the collider have, as a mass increase effect of 1:30,000 is the effect value.

So far, the biggest secret of nuclear weapons has been cracked. Its actual explosive force has been increased by 1,000 times in the atmosphere and 30,000 times in the light field!

CF. Dark Energy: The Biggest Mystery in the Universe

At the South Pole, astronomers try to unravel a force greater than gravity that will determine the fate of the cosmos
Richard Panek, April 2010

Scientists have some ideas about what dark matter might be—exotic and still hypothetical particles—but they have hardly a clue about dark energy.

Scientists reached a consensus in the 1970s that there was more to the universe than meets the eye. In computer simulations of our galaxy, the Milky Way, theorists found that the center would not hold—based on what we can see of it, **our galaxy doesn't have enough mass to keep everything in place. As it rotates, it should disintegrate, shedding stars and gas in every direction.** Either a spiral galaxy such as the Milky Way violates the laws of gravity, or the light emanating from it—from the vast glowing clouds of gas and the myriad stars—is an inaccurate indication of the galaxy's mass.

But what if some portion of a galaxy's mass didn't radiate light? If spiral galaxies contained enough of such mystery mass, then they might well be obeying the laws of gravity. Astronomers dubbed the invisible mass "**dark matter**."

The implication of that discovery was momentous: it meant that the dominant force in the evolution of the universe isn't gravity. It is...something else. Both teams announced their findings in 1998. Turner gave the "something" a nickname: **dark energy**.

Black Holes Are Mass-Energy and Energy-Mass Converters

Since Einstein proposed the theory of mass-energy conversion, people have gradually recognized a new cosmology, namely that the universe consists of two major elements, matter and energy, both of which can be converted to each other with the total amount remaining unchanged. However, in the way of conversion, this idea has not reached a unified understanding. In contrast, most people accept the theory of particle collision to achieve mass- energy conversion. The path of energy-mass conversion has always been in a confusing state.

This book unifies these two transformations and proposes a comprehensive and definitive solution, namely, the limit theory of the speed of light for particles. **The limit theory claims that; particles are energy when they travel at the speed of light, particles are matter when they move below the speed of light, and black holes are the most common converters.**

In the early stage of a black hole, vortex objects are whirled in, crushed, and accelerated to the speed of light to catalyse mass-energy conversion. At the later stage, with the increase of substance and the effect of continuously increasing pressure and tension, the rotating speed decreases and the energy-mass conversion process begins. Energy is converted into pure matter — hydrogen plasma - and the black hole is converted into a luminous celestial body.

This is the case with stars and their subclasses of small black holes. Galaxies and super-galactic black holes do not disappear until they are spiralled into larger black holes in their parent systems. They are powerful perpetual motion machines that constantly absorb objects pushed in by the spiral arms, tear them apart, dissolve them, and then eject them, forming new nebulae and stars that replenish the arms as new material. If the ejecta reaches the tail of the spiral arm, it forms a loop that connects the black hole from the top to the bottom. However, due to the long distance and numerous obstacles, only a small part of the ejecta can reach the tail of the spiral arms, and most of them are trapped by the spiral arms halfway, changing direction and returning to the black hole again. In addition, some of the ejecta hit the rotating arm material entering the black hole head-on and stop moving forward. Under the action of vortex motion, some elliptical celestial bodies form and gather around the black hole — or they participate in the construction of quasars, and then return to the black hole again.

The black hole eruption must be the result of the vortex contracting and accumulating material, which in a favourable scenario could reach a value equivalent to tens of billions of solar masses, creating a powerful and sustained jet stream that forms a new spiral arm. The arm is initially outward and diverged, but only when the jet ends and is constrained by the shell, do they turn toward the black hole. This is a new theory of spiral arm formation, which needs to be confirmed by future astronomical observations. **If this were true, each galaxy would become a relatively self-contained system in a super-galaxy, and the black hole a veritable cosmic metabolic machine.**

Nothing but Living Things Really Die

Not a ray is dimmed, not an atom worn; nature's oldest force is as good as new.
   Ralph Waldo Emerson: *Song of Nature*

In the universe, humankind, animals, and plants grow and die. Other natural things, such as water, earth, air, and various heavenly bodies, are always changing but experience no real death.

As discussed earlier, particles will not die, and now stars will not die either. One reason is that we cannot find their critical state or their corpses. It was originally thought that black holes were the corpses of stars, but now it has been proven that they are the foetal states of stars. Originally neutron stars, white dwarfs, red giants, and supernovae were considered to be in their critical state, but now they have been proven to be energetic stars or quasars.

In fact, **stars are made up of plasma and are the most dynamic things, never aging or dying but often changing.** At a higher level, galaxies and even the universe will not age and die, they will only change. Here we need to return to the difference between EBT and MBT.

According to MBT (matter-based theory), a star glows from energy generated by thermonuclear reactions, or nuclear fusion, within. The main raw material and fuel is hydrogen. This mechanism drives a series of chemical reactions. Hydrogen atoms are converted into lithium, helium, carbon, etc. This is a process of decreasing energy. Eventually, when all matter is converted into iron, no energy is produced, so the thermonuclear reaction or nuclear fusion ceases, the star enters a state of decay, ages and dies.

EBT denies the facts of thermonuclear reaction or nuclear fusion within stars and depicts a completely different picture. The cosmic background is an invisible energy field, that is, a light field. The most common motion is vortex motion. The most basic substance is hydrogen particles, which form visible nebulae. In the vortex motion, the nebula is rolled into a ball, forming a black hole. Then, it contracts further and hydrogen changes from a gaseous phase to a liquid plasma state, becoming a luminous star. This is a series of polymerization processes, in which matter and energy are converted to each other without any loss.

The body of the new-born star is liquid hydrogen, with a thick or thin surface of gaseous ions and a solid core. In general, a star is a plasma, a state in which ions are compressed, and must glow. The reason for this is that ions are protons or neutrons that move at a high speed without being bound by electrons. Like protons and neutrons, its energy (quarks, e.g., photons) is trapped and in a dispersed state; they are dull. When they are tightly packed together to form a plasma, collisions inevitably occur frequently, which destroy a small amount of ion shells, release photons, make stars visible, and even creating a fiery spectacle. The escaping photons make up starlight, also known as radiation, which is a loss to the star but only a negligible proportion. In addition, stars have multiple hot shells that strongly suppress the escape of free photons and even reflect them back, where they are trapped again and reduced to ions. The star is immortal.

Our Sun is an example of this. It makes balls of light and can also replenish energy from the light field without aging, due to luminescence. It has lost more than 90 percent of its angular momentum, and no longer contracts and explodes, much less collapses into a white dwarf. It is a pure plasma ball and will never change until the day it enters the black hole at the centre of the Milky Way.

Other matter in space — such as an isolated stone, a free molecule or an atom — will never die because **the vacuum can preserve everything perfectly as all energy trapped in things will never deteriorate.** Apart from living things, all changes can be attributed to external interactions, and the most important driving force is the ubiquitous vortex movement.

In general, vortex motion creates everything, including black holes, as it finally evolves everything into black holes and transforms them into new things. This is an endless cycle. For the universe, it is not a process of life and death but a process of eternal metabolism.

The Goal of the Grand Unification: The Unification of the Speed of Forces

"The only goal of science appeared to be analytical, i.e., the splitting up of reality into ever smaller units and the isolation of individual causal trains. Thus, physical reality was split up into mass points or atoms, the living organism into cells, behavior into reflexes, perception into punctual sensations, etc. Correspondingly, causality was essentially one-way: one sun attracts one planet in Newtonian mechanics, one gene in the fertilized ovum produces such and such inherited character, one sort of bacterium produces this or that disease, mental elements are lined up, like the beads in a string of pearls, by the law of association. Remember Kant's famous table of the categories which attempts to systematize the fundamental notions of classical science: it is symptomatic that the notions of interaction and of organization were only space-fillers or did not appear at all." (Bertalanffy, 1969, p.45)

We can add more 'one-way': The common ancestor of mankind is the East African, the vitality of all particles comes from the God particle, and the universe comes from the singularity explosion...

After the establishment of the theory of the four known forces, many scientists believe that they are only different manifestations of the same force, resulting in the grand unified theory (GUT) and the theory of everything (TOE). Force is motion, therefore the strength of every force is reflected by the speed of motion. As long as we find a way to make the speed of electromagnetic force, weak force, and strong force the same, the GUT can be proved. Its approach is high energy. In his book, Hawking gives the following explanation:

"The basic idea of GUTs (grand unified theories) is as follows: as was mentioned above, **the strong nuclear force gets weaker at high energies. On the other hand, the electromagnetic and weak forces, which are not asymptotically free, get stronger at high energies. At some very high energy, called the grand unification energy, these three forces would all have the same strength and so could just be different aspects of a single force.** The GUTs also predict that at this energy the different spin half mass particles, like quarks and electrons, would also all be essentially the same, thus achieving another unification." (2016, p.84)

There is no gravity here because it is too weak to be unified with the other three forces. However, many people are still trying to include gravity in this theory called the theory of everything, or, TOE. Einstein's unified field theory belongs in this class.

Following this line of thinking, Steven Weinberg did not use grand unification energy and discovered weak force messenger particles, namely, the W particle and the symmetrical Z particle. The two particles unify the weak force and the electromagnetic force. The next step, naturally, is the unification of these two forces and the strong nuclear force, only to find that an insurmountable obstacle is also high energy: a machine that is powerful enough to accelerate particles to the grand unification energy would have to be built as big as the solar system. (Hawking, 2016, p.85)

Returning to Hawking's previous example of a roulette game. If we select 37 men and women to form a mixed sports team, **give stimulant to all of the weaker players, and require strong members to slow down so that their running speed can reach unity,** would this be considered the goal of grand unification? Can the essential differences between the stronger players and the weaker players (strong force and weak force) be eliminated in this way?

The difficulty with GUT is that it needs to find a lot of messenger particles. Going back to the example above, it would take more coaches, nutritionists, body doubles, etc., to get the 37 athletes at the same pace.

"If the strong field is to be brought under the GUTs umbrella, it would seem that **there must be ultra-massive exchange particles required beyond the gluons, no less than twelve of them. Because they are more massive than the W's and Z's, they will be harder to detect, and there is no hope for them right now.** They will also be far shorter in range than anything that has yet been considered. The range of action of these ultra-massive exchange particles of the strong field is less than 1 quadrillionth the diameter of the atomic nucleus." (Asimov, 1993)

Asimov went on to talk about the reasons why scientists are obsessed with GUT and the achievements they have made, although gravity is unlikely to be unified.

"The theories involved in these new unification can be used to work out the details of the big bang with which the universe started. It would seem that at the very start, when the universe had existed for less than a millionth of a trillionth of a trillionth of a trillionth of a second and was far tinier than a proton and had a temperature in the trillions of trillions of trillions of degrees, there was only one field and only one kind of particle interaction. As the universe expanded, and the temperature dropped, the different fields 'froze out'.

Thus we could imagine the earth, if extremely hot, to be nothing but a gaseous sphere in which all the different kinds of atoms would be evenly mixed so that every portion of the gas would have the same properties as every other.

As the gas cooled, however, different substances would separate out first as liquids, then as solids; and eventually there would be a sphere of many different substances existing separately.

So far, though, the gravitational interaction proves intransigent. There seems no way of including it under

the umbrella of the kind of mathematics worked out by Weinberg and the rest. The unification that defeated Einstein has so far defeated all his successors as well.

Even so, the GUTs has produced something extremely interesting, indeed. Physicists have wondered how the big bang could produce a universe so lumpy as to have galaxies and stars. Why did not everything simply spread out into a vast haze of gas and dust in all directions? Then, too, why is the universe of such a density that we cannot be quite certain whether it is open or closed? It might have been distinctly open (negatively curved) or closed (positively curved). Instead, it is nearly flat." (Asimov, 1993)

Although *Force* is a common term, our understanding of it is uncertain and there is no unified definition. The following three paragraphs in Hegel's *Shorter Logic* (§136) will help us to deepen our discussion.

"Force consequently requires solicitation from without: it works blindly: and on account of this defectiveness of form, the content is also limited and accidental. It is not yet genuinely identical with the form: not yet is it as a notion and an end; that is to say, it is not intrinsically and actually determinate. This difference is most vital, but not easy to apprehend: it will assume a clearer formulation when we reach Design. If it be overlooked, it leads to the confusion of conceiving God as Force, a confusion from which Herder's God especially suffers...

whereas force is only shown to be force when it exerts itself, and in its exercise only comes back to itself. The exercise is only force once more. Yet, on further examination even this relation will appear finite, and finite in virtue of this mediation: just as, conversely, the relation of whole and parts is obviously finite in virtue of its immediacy. The first and simplest evidence for the finitude of the mediated relation of force and its exercise is, that each and every force is conditioned and requires something else than itself for its subsistence. For instance, a special vehicle of magnetic force, as is well known, is iron...

Hence in empirical physics, we speak of the forces of gravity, magnetism, electricity, etc., and in psychology of the forces of memory, imagination, will, and all the other faculties. All this multiplicity again excites a craving to know these different forces as a single whole, nor would this craving be appeased even if the several forces were traced back to one common primary force. Such a primary force would be really no more than an empty abstraction, with as little content as the abstract thing-in-itself. And besides this,

the correlation of force and manifestation is essentially a mediated correlation (of reciprocal dependence), and it must therefore contradict the notion of force to view it as primary or resting on itself."

HIGHLIGHTS

Force 'is not intrinsically and actually determinate.'
'Force is only shown to be force when it exerts itself.'
'Every force is conditioned and requires something else than itself for its subsistence.'
'Such a primary force would be really no more than an empty abstraction.'

A piece of iron rests on the table, it's not showing any force. When we put it on the scale, it shows gravity. Only when I touched it with another piece of iron did I know whether it had magnetic force or not. This is what Hegel said 'Force is only shown to be force when it exerts itself.'

One fact confirmed by contemporary scientific observation is that vortex motion is everywhere, and they are the greatest forces. Compared with it, the four forces are insignificant and can be ignored. Another fact confirmed by scientific observation is that material vibration is eternal, even if stuffs stop moving. This fact denies the possibility of unification of the four forces for one reason: the vibration wavelengths and frequencies of each force are extremely different and cannot be unified.

Scientists Have Been Exploring Electricity and Magnetism for 400 Years

When discussing the discovery of electricity, people will think of the phenomenon of triboelectrification, and go back to the Greek Thales of Miletus more than 2500 years ago.

Amber ornaments are ancient Greeks' common handicraft products, which they often use to make love charms. They found that rubbing amber can attract feathers, threads and other small things, so they called this phenomenon elektron, meaning 'glowing sun', which was written as electrum in Latin and evolved into electricity in English. This is also a magnetic phenomenon. The Greeks discovered both electricity and magnetism.

This discovery had not received extensive attention for a long period, over 20 centuries, until 1600 when **William Gilbert,** the most distinguished man of science in England during the reign of Queen Elizabeth, and the father of electric and magnetic science, proposed in his *De Magnete* in 1600 that electricity and magnetism, while both capable of causing attraction and repulsion of objects, were distinct effects. But later researchers generally ignored or did not agree with his view.

Magnetic poles and electric charges at one time seemed rather similar: in each case, two kinds existed, opposites attracted and likes repelled. As French physicist **Charles Augustin de Coulomb** found around 1777, the forces between charges or poles always diminished with distance r like $1/r^2$, which was also the way gravity decreased. They are considered to have internal connection, although there are external differences between them.

This connection was first proposed in 1920 by Danish physicist **Hans Christian Oersted.** He showed the connection between magnetism and electric current when a magnetic field produced by a current-carrying copper bar deflected a magnetised needle during a lecture demonstration, he thought that magnetic effects radiate from all sides of a wire carrying an electric current, as do light and heat. firstly to identify the force of electromagnetism.

Afterward, the theory of electromagnetism, known as classical electromagnetism, was developed by various physicists between 1820 and 1873 when it culminated in the publication of a treatise by **James Clerk Maxwell**, who unified the preceding developments into a single theory and discovered the electromagnetic nature of light. In classical electromagnetism, the behaviour of the electromagnetic field is described by a set of equations known as Maxwell's equations, and the electromagnetic force is given by the Lorentz force law.

One of the peculiarities of classical electromagnetism is that it is difficult to reconcile with classical mechanics. According to Maxwell's equations, the speed of light in a vacuum is a universal constant that is dependent only on the electrical permittivity and magnetic permeability of free space. This violates Galilean invariance, a long-standing cornerstone of classical mechanics. One way to reconcile the two theories (electromagnetism and classical mechanics) is to **assume the existence of a luminiferous ether**, through which the light propagates.

However, subsequent experimental efforts failed to detect the presence of the ether. After important contributions of Hendrik Lorentz and Henri Poincaré, in 1905, Albert Einstein solved the problem with the introduction of special relativity, which replaced classical kinematics with a new theory of kinematics compatible with classical electromagnetism.

In addition, relativity theory implies that, in moving frames of reference, a magnetic field transforms to a field with a nonzero electric component and conversely, a moving electric field transforms to a nonzero magnetic component, thus firmly showing that the phenomena are two sides of the same coin. Hence the term electromagnetism.

Maxwell's equations have been superseded by quantum electrodynamics (QED). The electromagnetic field is quantized by imagining that at every point in space and time is a quantum harmonic oscillator. The empty field (vacuum state) fluctuates randomly as a consequence of the uncertainty principle.

How does the electromagnetic force work? Stephen Hawking offers the answer. **"The electromagnetic attraction is pictured as being caused by the exchange of large numbers of virtual massless particles of spin 1, called photons. Again, the photons that are exchanged are virtual particles."** (2016, p.80) "the electric repulsive force between two electrons is due to the exchange of virtual photons." (p.78)

Here we can see that **virtual photon is like a smart arbitrator.** It knows to pull some kind of particles together and push the other kind away to prevent them from impinging with each other. How amazing!

And in that case, each atomic family has to expand greatly. Say, a gold atom, according to the particle theory, besides having 79 protons, 79 electrons and 118 neutrons, it has to 'employ' at least 197 gluons as 'nanny' to hold the quarks in each proton and neutron tightly and keep them from being scattered; 'hire' dozens of stronger gluons as 'doormen' to hold all protons and neutrons in the nucleus. In addition, 79 virtual photon 'bodyguards' have to be 'employed' to prevent each electron from being collided by other electrons or colliding with other electrons. How do these hundreds of members live in harmony in a tiny atomic space? In my eye, it can't be anything but a mess !

One of people's basic expectations for scientists is to explain abstract vague phenomena and concepts with concrete and clear concepts, but scientists do the opposite, and explain Greek clear phenomena and concepts with vague and even mysterious terms such as electromagnetism, oscillator, virtual photon and virtual particles, which is quite disappointing to the public.

Electricity, A Common Widespread Misconception

Today, the power grid is all over the world, electricity has become the main power for people daily life, most people can not leave electricity for a moment. Yet to this day, the concept of electricity remains vague, contradictory, and continually debated.

What's electricity? I found a variety of answers from the Internet.

"1. Any phenomenon associated with stationary or moving electrons, ions, or other charged particles. 2.an electric current or charge: a motor powered by electricity." *Collins English Dictionary*

"1. A fundamental property of matter caused by the presence and motion of electrons, protons, or positrons, manifesting itself as attraction, repulsion, luminous and heating effects, and the like. 2. electric current or power." *Random House Kernerman Webster's College Dictionary*

"1. The collection of physical effects resulting from the existence of charged particles, especially electrons and protons, and their interactions. Particles with like charges repel each other. Particles with opposite charges attract each other. 2. **The electric current generated by the flow of electrons around a circuit and used as a source of power.**" *The American Heritage® Student Science Dictionary*

"Electricity, phenomenon associated with stationary or moving electric charges. Electric charge is a fundamental property of matter and is borne by elementary particles. In electricity the particle involved is the electron, which carries a charge designated, by convention, as negative. Thus, the various manifestations of **electricity are the result of the accumulation or motion of numbers of electrons.**" (Encyclopedia Britannica)

"**Electricity is the term used to define the**

phenomenon where electrons move from one atom to another, producing an electrical charge... Electricity is when electrons move from one atom to another, in much the same way the ping-pong balls were passed from one person in the circle to another. The flow of electricity is called a current, which we measure in amperes."
(https://study.com/academy/lesson/)

As mentioned above, reductionism is a bottom-up approach that focuses on particles & points and is good at establishing complex calculation formulas and mathematical models on this basis, while the definition and nature of research objects are always unclear. The same is true of gravity, of corpuscles, of quantum theory, of magnetism & electricity. My approach, contrary to reductionism, is top-down. In order to understand the properties of magnetism & electricity, let me introduce two notions, and compare them.

First, **Tony R. Kuphaldt's 'potential energy imbalance pushes electron flows for the requirement of resistance' notion.** He is an instructor in the Instrumentation and Control Technology Department at Bellingham Technical College, Washington Area. It's classic and adopted by most schools at present. Here I quote his popular online works, *Lessons In Electric Circuits*, copyright (C) 2000-2020, under the terms and conditions of the CC BY License, Revised November 06, 2021.

"The result of an imbalance of this 'fluid' (electrons) between objects is called static electricity...By the time the true nature of electric 'fluid' was discovered, Franklin's nomenclature of electric charge was too well established to be easily changed, and so it remains to this day."

"While the normal motion of 'free' electrons in a conductor is random, with no particular direction or speed, electrons can be influenced to move in a coordinated fashion through a conductive material. This uniform motion of electrons is what we call electricity, or **electric current.**

To be more precise, it could be called dynamic electricity in contrast to static electricity, which is an unmoving accumulation of electric charge."

"An electric circuit is formed when a conductive path is created to allow free electrons to continuously move. **This continuous movement of free electrons through the conductors of a circuit is called a current, and it is often referred to in terms of 'flow', just like the flow of a liquid through a hollow pipe."**

"**The force motivating electrons to 'flow' in a circuit is called voltage. Voltage is a specific measure of potential energy that is always relative between two points.** When we speak of a certain amount of voltage being present in a circuit, we are referring to the measurement of how much potential energy exists to move electrons from one particular point in that circuit to another particular point. Without reference to two particular points, the term 'voltage' has no meaning."

"Free electrons tend to move through conductors with some degree of friction, or opposition to motion. This opposition to motion is more properly called resistance. The amount of current in a circuit depends on the amount of voltage available to motivate the electrons, and also the amount of resistance in the circuit to oppose electron flow. Just like voltage, resistance is a quantity relative between two points. For this reason, the quantities of voltage and resistance are often stated as being 'between' or 'across' two points in a circuit...The Galvanic Circuit Investigated Mathematically. **Ohm's principal discovery was that the amount of electric current through a metal conductor in a circuit is directly proportional to the voltage impressed across it,** for any given temperature. Ohm expressed his discovery in the form of a simple equation, describing how voltage, current, and resistance interrelate: **$E=IR$.** In this algebraic expression, voltage (E) is equal to current (I) multiplied by resistance (R)."

"$P=IE$... It must be understood that neither voltage nor

current by themselves constitute power. Rather, power is the combination of both voltage and current in a circuit."

REVIEW

Kuphaldt offers a perpetual motion model: voltage-current-resistance-voltage, a closed loop works by potential energy. The disadvantage of this model is that the cause of potential energy generation is unknown. Potential energy is not inherent in matter, it is energy converted from another kind of energy. Triboelectrification is a potential energy artificially created, which is static electricity, and dynamic electricity cannot be generated in this way. Reminder. In CFT, free electrons are energy particles and are self-driven.

Second, **William J. Beaty.** Office webcam, UW campus, Research Engineer, University of Washington. **His view is unique and subversive.** I'm quoting his work that was posted online, *Misconceptions Spread by K-6 Textbooks: 'Electricity'*, 1995. (http://amasci.com/miscon/elect.html)

Here are his opening remarks and main ideas.

"How does 'electricity' work? If you've learned about electricity from grade-school textbooks, then **first we have to do some 'debunking' and find out how electricity DOESN'T work.** Sorry if the following is a bit contentious at times. I wrote it in an attempt to get some things off my chest. If you keep watching this site, I'll probably clean it up and make it sound a bit more professional. Also, this file is still under construction and is being written, edited, corrected, etc. It does currently contain some mistakes of its own. I placed it online as a sort of 'trial by fire' in order to hear readers' responses and target weak or unclear sections for improvement. Please feel free to post comments. -Bill B."

"Electrons and protons are matter, not energy. **A flow of**

electrons is NOT a flow of energy, it is a flow of matter and a flow of electric charge. Same goes for protons."

"The truth is that the word 'Electricity' has many contradictory meanings and so the word itself has become meaningless. **Electricity is not an event. Neither is it energy, or electrons, or electron motion. Electricity is just a big mistake, but a mistake that slowly crept up on everyone.** We never realized it was happening. As long as we keep trying to figure out what 'electricity' really is, we'll keep spreading the confusion. The only honest move is to stop hiding the problem. Stop the cover up. We should perform an act of painful honesty, and admit that we've been accidentally misleading generation after generation of students by teaching them about the wonderful substance/occurrence/energy called 'electricity' ...which doesn't really exist."

"DURING A CURRENT, ELECTRONS IN WIRES START JUMPING FROM ATOM TO ATOM? Wrong.

When individual atoms of copper are brought together to form a bulk metal material, something unexpected happens. **The outer electron of each copper atom leaves its parent atom. Rather than orbiting single atoms, the outer electrons all begin 'orbiting' around and among ALL the atoms in the metal.** Essentially the metal's electrons are 'jumping' from atom to atom all the time, even when there is no electric current applied. Physicists call this the 'electron sea' or 'electron gas' of the metal.

Rather than jumping all the time, what would happen if the electrons only jumped between atoms during electric currents? Well, their jumping motion gives wires their conductivity. If the electrons jumped less during smaller currents, and they stopped jumping during zero current, then the metal conductivity wouldn't be constant, and Ohm's law wouldn't apply. Instead, conductivity would decrease as current decreased. (We know that this doesn't happen.) And if the electrons stopped jumping entirely, the metal would become an insulator. Does copper become insulating during low or zero current? Of course not. **So, it's wrong to believe that electric current causes electrons to start hopping. Instead, their constant built-in 'hopping' gives conductors a particular value of conductivity. The constant hopping also makes the population of electrons behave like a movable fluid. That's what makes wires so wonderful: they act like pre-filled pipes. They're filled with 'liquid electrons'."**

"BATTERIES STORE CHARGE, AND THIS CHARGE FLOWS IN WIRES? No.

The word 'charge' has more than one meaning, and the meanings contradict each other. **The 'charge' in a battery is energy (chemical energy), while the 'charge' that flows inside wires is part of matter, it is electron particles.** And those wires, even though full of charge... are neutral and uncharged! The term 'charge' refers to several different things: to net-charge, to quantities of charged particles, and to 'charges' of energy. If you are not very careful while using the word 'charge' in teaching, you might be spreading misconceptions. For example, even when metals are totally neutral, they contain vast quantities of movable electrons, immense charge. So, should we say that they contain zero charge because they are neutral?"

"ELECTRIC POWER FLOWS FROM GENERATOR TO CONSUMER? Wrong.

Electric power cannot be made to flow. Power is defined as 'flow of energy'. **Saying that power 'flows' is silly. It's as silly as saying that the stuff in a moving river is named 'current' rather than named 'water'.** Water is real, water can flow, flows of water are called currents, but we should never make the mistake of believing that water's motion is a type of substance. Talking of 'current' which 'flows' confuses everyone. The issue with energy is similar. Electrical energy is real, it is sort of like a stuff, and it can flow along. When electric energy flows, the flow is called 'electric power'. But electric power has no existence of its own. Electric power is the flow rate of another thing; **electric power is an energy current. Energy flows, but power never does, just as water flows but 'water current' never does.** The above issue affects the concepts behind the units of electrical measurement. Energy can be measured in Joules or Ergs. The rate of flow of energy is called Joules per second. For convenience, we give the name 'power' to this Joule/sec rate of flow, and we measure it in terms of Watts. This makes for convenient calculations. Yet Watts have no physical, substance-like existence. The Joule is the fundamental unit, and the Watt is a unit of convenience which means 'joule per second.'"

"ELECTROMAGNET COILS USE UP ENERGY TO MAKE MAGNETISM? Not right.
  Sustaining a magnetic field requires no energy. Coils only require energy to initially create a magnetic field. They also require energy to defeat electrical friction (resistance); to keep the charges from slowing down as they flow in wires. But if the resistance is removed, the magnetic field can exist continuously without any energy input. If electrically frictionless superconductive wire is used, a coil can be momentarily connected to an energy supply to create the field. **Afterwards the power supply can be removed and both the current and the magnetic field will continue forever without further energy input.**"

"ATOMS HAVE EQUAL NUMBERS OF ELECTRONS AND PROTONS? Not in conductors!
  Many students misunderstand how electric circuits work. One reason for this is that they think the electrons in a metal are trapped on individual metal atoms. They also think that an applied voltage is needed to 'free' the electrons and to change metal into a conductor. They aren't aware that the 'sea' of free electrons always exists inside metal all the time. I suspect that this is part of a more general misconception that all atoms in a material are always neutral. This is wrong because **all conductors contain charged, movable**

particles. The very definition of 'conductor' is 'a material which contains mobile charges'. If all atoms were truly neutral, then conductors could not exist."

"ELECTRIC ENERGY TRAVELS INSIDE OF WIRES? Wrong.

When an electric company's distant generator lights up your lamps, the electric energy travels along the power lines at almost the speed of light. Most K-12 textbooks teach that the energy is trapped inside of electrons, and these electrons flow inside the wires. Doesn't this mean that electrical energy flows INSIDE the metal wires? Nope, since electrical energy is not trapped inside electrons. Instead **the energy is made of invisible magnetic fields and electric fields which surround the electrons, and these fields surround the wires. Electrons don't flow fast like the energy does, instead they ooze along slowly to produce an electric current.** But how can electrons flow slowly if the energy flows fast? It's because the energy can leap from electron to electron. Indeed, the energy is connected to a whole vast population of electrons in the wire, and it isn't attached to any single one.

Is this confusing? Here's another way to see it. During an electric current, the wires become surrounded with magnetic field. This field IS the electrical energy. Also, whenever a pair of wires is connected to a battery or generator, the two wires become oppositely charged, and they become surrounded with an invisible electrostatic field. **This field IS the electrical energy. Magnetic and electric fields exist in the empty space surrounding your lamp cord, and these fields contain the 'wattage', they contain the flow of electrical energy that powers the light bulb. Electric and magnetic fields together are 'Electromagnetism', the same kind of energy as radio waves and light.** Those 'EMFs' that people worry about; the invisible 'EMFs' that surround wires and exist invisibly in our homes... that's the electromagnetic energy which lights our lights and runs our appliances. It certainly makes sense that it travels at the same speed as radio waves and light, since it's made of the exact same 'stuff'. So, what would electrical energy look like if we could see it? Here are some simple drawings I made. Whenever you plug in a light bulb, the energy that flows along the lamp cord is like a fuzzy sausage a couple of inches thick. It follows the two wires of the cord, then it dives into the thin filament of the light bulb. Is this a very strange description of 'electricity?' Yes and no. It's the same description taught to advanced engineers in their courses on waveguides and radio physics. It's also taught to university physics students, especially if they read chapter 27 of The Feynman Lectures. But it's not so terribly advanced, instead it's just unfamiliar, because so few people know about it or discuss it."

SUMMARY

'Electricity' is a common widespread misconception. Actually, it does not exist.

Electrons & protons are matter, not energy. A flow of electrons is NOT a flow of energy, it is a flow of matter and a flow of electric charge... Electrons don't flow fast like the energy does, instead they ooze along slowly to produce an electric current.

This field IS the electrical energy. Magnetic and electric fields exist in the empty space surrounding your lamp cord, and these fields contain the 'wattage', they contain the flow of electrical energy that powers the light bulb. Electric and magnetic fields together are 'Electromagnetism', the same kind of energy as radio waves and light.

That's the electromagnetic energy which lights our lights and runs our appliances. It certainly makes sense that it travels at the same speed as radio waves and light.

That's right. **The energy of electricity comes from a magnetic field. The further discussion is, what's it? How does it work?**

The Simplest and Clearest Principles of Electricity & Magnetism

From the above discussion, I find that modern electromagnetic science has moved far away from the Greek discovery, and that electricity and magnetism are very intertwined. In my opinion, to know the truth about electricity and magnetism, we must go back to the Greeks and start anew.

Common sense tells us that glass rods and rubber rods do not conduct electricity, so they are called insulators. Why are they charged after rubbing? Therefore, some people changed triboelectrification to friction magnetism. From the previous discussion, these two statements are correct. Friction insulators generate both magnetism & static electricity, they're actually one. Glass rod and rubber rod are composed of their own atoms and molecules. They do not flow but random vibration, so they have no magnetism or electricity. Friction makes some molecules and atoms lose electrons and become cations, and makes other part of molecules and atoms get extra electrons and become anions. Their vibration is enhanced and move in opposite directions, but they cannot move. forming a potential difference. Thus magnetism (cation force) & static electricity are created. Rubbing copper bars and iron bars has no such effect. This is because they are good conductors, and the generated potential energy and potential difference will be eliminated soon. According to this common characteristic, I give the simplest and clearest description of magnetism & static electricity:

The net charge of atoms in an object is non-zero due to its total number of electrons being unequal to its total number of protons, total atoms become electrolytes, is magnetism; Electrolytes resonant orderly is static electricity.

NOTE. Although ions account for 99% of the total amount of matter, most people are still very confused about it. **The most obvious thing is to confuse cation-anion(electrolyte) with ion.** In fact, the real ions are bare protons. They often vibrate violently and glow. Electrolytes, by contrast, vibrate gently and generally do not emit light. Both sunlight & electricity are typical plasma resonance.

Obviously, cation & anion are actually free-swimming **ATOMS, not IONS.** Their energy is confined, their movement depends on external energy, and their speed is slow. The energy of static electricity, voltaic pile and triboelectrification results from the resonance of such atoms. **I call this static or weak ionization, and the other dynamic or strong ionization. They produce static and dynamic electricity respectively.**

Here, static electricity is magnetism, cation field, while dynamic electricity is ion field. They are not what people call electricity, hence Beaty's critique on electricity, 'a common widespread misconception.'

The two sorts of electricity can induce each other. Oersted demonstrates the fact that dynamic electricity induces static electricity, the magnetism. **I named this interaction atom-ion tautometry, or, A-I Cycle. Our power system allows us to get this cheap energy in this way.**

"It could be called dynamic electricity in contrast to static electricity, which is an unmoving accumulation of electric charge." (Kuphaldt) Naturally, the way to store dynamic electricity is to convert it to static electricity, though inefficient. Car battery is made on this principle.

The founder of the static ionization was Joseph John Thomson. "The honor of making the first incision in the complicated operation of dissecting the delicate body of the atom belongs to the famous British physicist J. J. Thomson, who was able to show that the atoms of various chemical elements consist of positively and negatively charged parts, held together by the forces of electrical attraction. Thomson conceived an atom as a more or less uniformly distributed positive electric charge with a large number of negatively charged particles floating in its interior The combined, electric charge of negative particles, or of electrons as he called them, equals the total positive charge, so that the atom on the whole is electrically neutral. Since, however, electrons were assumed to be bound comparatively loosely to the body of the atom, one or several of them could be removed leaving behind a positively charged atomic residue known as positive ions. On the other hand, the atoms that manage to get into their structure several extra electrons from the outside have an excess of negative charge and are known as negative ions. The process of communicating to an atom a positive or negative excess of electricity is known as the process of ionization." (Gamow,1947, p.131-132)

Combing hair is actually an act of making cations & anions, the weak ionization. If a hair is used as a filament and electrified (as Thomas Edison did), strong ionization will occur. All atoms of the hair will ionize, heat up, glow, and will soon be burned out. If a tungsten filament is replaced hair, it will also burn out, but it takes much longer time to glow.

A circuit is composed of a pure metal conductor, which is in the off state at first, and has neither electricity nor magnetism. Once it is connected into a magnetic field, all electrons in the conductor will form a unified electric field, and all bare protons will form an ion field. **In A-I Cycle, the vibration of the ion field is both resistance and force, and the three fields interact. The relationship between them can be described by Ohm's law, U = IR. That is, a magnetic field(U) = an electric field(I) × an ion field(R).**

From this view, the electricity in a power system is the force of the A-I Cycle. Its frequency is usually 50-60 cycles per second, and its unit is abbreviated to HZ.

This model completely revolutionized the classic concept of electricity. It replaces 'charge flow' or 'electron flow' with electric field, and recognizes that circuit is the unity of opposites between the negative

electron field and the positive ion field; dynamic electricity is the ionic resonant vibration force.

From this definition, we would not take it for granted that the task of a power station is to supply 'ionic resonant vibration flow' to users. Nope, Nature pursues the shortest way, the highest efficiency, and does not do such awkward things. The reality is that the power station only provides a quick-intermittent rated magnetic field (volt) and an electric field induced by it. The core of all electrical appliances is an independent closed conductor. When the user connects one to the power system, its core conductors is instantly ionized and runs in sync with the system A-I Cycle, and its resonant vibration force allows electrical appliance to push the ionic vibration of electrical appliance, such as light an electric bulb or a television set; Push molecules of stuffs to vibrate, as in microwave ovens; Or by converting parts to drive an electric machine, etc.

The power station produces artificial electricity, its magnetic field (voltage) and electric field (power) are certain, and its ion field is determined by the user's electricity consumption, is changeable. Dynamic electricity can't be stored, which leads to the phenomenon that the brightness of electric lamps fluctuates during the peak & low periods of electricity consumption.

● Concerning the principle of the circuit.

The system circuit ion field is excited by resonance with consumer appliances, that in turn induces the system circuit magnetic field, that in turn induces the dynamic electric field, which resonates with consumer appliances again to produce ion field, hence a harmonious A-I cycle. If no electrical appliance is connected, all atoms of the system circuit will be converted into electrolyte, only magnetic field and electrostatic field (cation field), which will generate a slight heat, but no work will be done. It's like a circuit made up of a battery connected by a copper wire. The copper wire will become an electrolyte, which will vibrate a little faster, heat slightly, but won't emit light. Only when a small bulb is connected to trigger ion resonance, which generates dynamic electricity, and light up. In other words, the power station and the customer are connected by one electric circuit all the time, together, they make the system work.

There's a passage in Engels's *1883-Dialectics of Nature,* that helps us understand circuit.

"An unconnected, superfluous metal in the battery acts as a non-conductor; it can neither form ions nor allow them to pass through, and **without ions we know of no conduction in electrolytes.**"

Now we can think of it as, the wires and the metal parts of the system circuit are all electrolytes, 'an unconnected, superfluous metal', '**non-conductor**', and only the metal with higher resistance (such as a tiny piece of tungsten wire) that are connected to the user's electrical appliances, which make all atoms of the metal convert into ions by the system's magnetic field (voltage), are working as **conductors**.

Magnetic Field-Electric Field-Ion Field — the Triad of Electricity

A consensus in the physics community is that Michael Faraday and James Clerk Maxwell established the electromagnetic field theory.

"Every particle possessing electric charge is the source of an electromagnetic field that stretches outward in all directions indefinitely, the intensity of the field decreasing in proportion to the square of the distance from the source. Every particle possessing both mass and electric charge (and there is no electric charge without mass) is the source of both fields... In the 1860s, Clerk Maxwell, a great admirer of Faraday, set about supplying the mathematical analysis to account for the lines of force. In doing so, he evolved a set of four simple equations that among them described almost all phenomena involving electricity and magnetism. **These equations, advanced in 1864, not only described the interrelationship of the phenomena of electricity and magnetism, but showed the two cannot be separated.** Where an electric field exists, there has to be a magnetic field, too, at right angles; and vice versa. There is, in fact, only a single electromagnetic field." (Asimov, 1993)

In R.A. Brooks' book, we see the nature of this theory and its huge social impact, which inspired Max Planck to establish quantum theory.

"The electromagnetic field, like gravity, is a force field; it is the agent that transmits force from one electric charge to another. It consists of two component fields, electric and magnetic. **Unlike gravity, which is always attractive, the EM force can be attractive, repulsive, or even sideways.** Its field nature was suggested by Michael Faraday in 1845 and field equations were developed by James Maxwell in 1864. These equations predicted EM waves (oscillations in field intensity) that travel at a speed of 300,000 km/second, which turned out to be the same as the speed of light. Not only did Maxwell's equations thus explain the entire field of optics, they also predicted a vast spectrum of EM radiation that includes radio and TV waves, microwaves, infrared, visible light, ultraviolet, x-rays, and gamma rays.

The quantum nature of the EM field was discovered by Max Planck while studying radiation from a hot object. He found that **the data made sense if the radiation is not infinitely divisible, but consists of discrete amounts of energy that he called quanta. These quanta are now called photons. Planck also found that the energy of a photon is proportional to its frequency of oscillation."** (Brooks, 2016, p.70)

In this way, Maxwell united electricity, magnetism, and light into the electromagnetic spectrum. However, they are three different physical facts, how do they interact with each other, and become one?

"It would seem by **Heisenberg's theory that all forces of attraction and repulsion would be the result of exchange particles.** In the case of electromagnetic attraction and repulsion, the exchange particle is the photon." (Asimov, 1993)

Here, the photon (virtual photon) is the main actor, the electromagnetic field is the stage, it is not the answer we are looking for. To understand the relationship between electricity and magnetism, we let the photons go and focus on the electromagnetic field only.

Brian Clegg, an award-winning pop science writer, describes it in his book, *Gravitational Waves: How Einstein's Spacetime Ripples Reveal the Secrets of the Universe*, *electricity and magnetism interact in Maxwell's electromagnetic field.*

"The remarkable Scottish physicist James Clerk Maxwell

worked out in the early 1860s that light was an interaction between electricity and magnetism. And this meant that in principle, you could have **an electric wave creating a magnetic wave, creating an electric wave and so on**, hauling itself through empty space by its own bootstraps without any material medium required —it is the electromagnetic field that acts as the material." (2018)

That's what Maxwell meant. But this is the interaction of two fields, not the motion of a unified electromagnetic wave —does electromagnetic wave exist?

"Where an electric field exists, there has to be a magnetic field, too, at right angles; and vice versa. **There is, in fact, only a single electromagnetic field.** ..In considering the implications of his equations, Maxwell found that a changing electric field has to induce a changing magnetic field, which in turn has to induce a changing electric field, and so on; the two leapfrog, so to speak, and the field progresses outward in all directions." (Asimov, 1993)

I mean, **magnetic waves don't have a place in the electromagnetic spectrum.** Light waves and radio waves, electric fields and magnetic fields are well understood, but magnetic waves are poorly known. So here we have velocity and direction of motion. Light, electricity, and their fields all move at the speed of light and are transmitted outward. Magnets and magnetic fields also move, but only toward the center, then fixed. It doesn't spread out like a light field, it doesn't send out waves like an electric field.

In the previous discussion we determined that an electrical system consists of three fields which are interrelated and interacting with each other. **The magnetic field (positive cation field) induces an electric field (negative electric field), which induces an ion field (positive electric field), and the ions resonance in turn produces a magnetic field, which is a closed loop.**

For ease of understanding, we imagine a separated

family of three (three fields in one), with the wife (electron) and baby (proton) imprisoned in a glass house. She held the babies close all the time and they were in a stationary state (atoms). When her husband (magnet) approached, she immediately put down the baby and went to her husband, the baby (ion) instantly jumped to be held by her mother—the connection of the circuit. When she separated from her husband, and hugged the baby, they returned to stationary again—the circuit was disconnected. If the husband and wife meet on a regular and recurring basis, it's alternating current. I dub it A-I Cycle.

In the theory of electromagnetic field, there are only husband and wife, no baby; only deformed cosmic energy (virtual photons) and no material energy. It is mathematically possible, but practically impossible — Electricity cannot be generated in an electromagnetic field by exchanging charges or virtual photons.

The new electric mode is not only suitable for daily life, but also for celestial evolution. Nebulae are mixtures of photons, ions, atoms, molecules and objects, and the sun evolved from the nebula vortex. The vortex and the subsequent black hole spin violently, forming a powerful magnetic field, which induces a strong electric field & ion field. In this way, the sun becomes a plasma, a luminous body. The sun is a one-off electric mode, with only A-I conversion and no I-A conversion. The same as the other stars. The one-way I-A conversion occur on planets.

Typical of the magnetic-electric-ionic trinity is our electric power system, but not all electricity is like this. According to the new definition, electricity (ion field) can be generated by any force which ionizes atomic stuffs. Such as battery, lightning, thermoelectricity, ionosphere of the earth, and so on. They can induce magnetic field, but magnetic field is not necessary for these kinds of electricity.

As an end product, electricity is an electric field & an ion field, which constitute negative and positive charges respectively. It is the positive charge that does all works, spurring ion resonance both the basic ion field of the system and fields of the external electrical appliances. Differ from the popular view, methinks the negative electricity neither does work nor flows, and its function is just to maintain all ion fields on A-I cycles.

CONCLUSION

Electricity consists of three fields, and only field theory can explain electric phenomena, particle theory can't. Magnetism is static electricity, although it and dynamic electricity are in the same power system, mutual influence, mutual conversion, and mutual induction, but are independent of two fields, can not be integrated into one. That is, the unified electromagnetic force & field is untrue. We can use mathematics to establish it, as it doesn't exist in reality. Like an atom, it has a weak electron field and a strong nucleon field, and each atom has those two fields, but they don't merge into one field.

LFT vs Relativity

Albert Einstein had said that there were only twelve people living who understood his theory of relativity although a good many books had been written to explain it.

I think even today, more than a century after the publication of the theory of relativity, there are not more than a dozen people who really understand it. Here again is the superiority of holism.

An indispensable job for anyone to establish cosmology is to establish a space-time frame, in which all objects are located and all events occur, viewed as a single and continuous framework for existence. It reflects people's general recognition of the cosmos.

The Greeks established ether theory, which is also the default spacetime frame of Descartes and Maxwell. It has always been an ambiguous and imperfect frame.

Newton established absolute frames of spacetime also on the basis of ether, which are described as: "I. Absolute, true, and mathematical time, of itself, and from its own nature flows equably without regard to anything external...II. Absolute space, in its own nature, without regard to anything external." (*The Principia*, 7) Therefore, Newton doesn't have to consider the influence of spacetime when discussing various motions and forces.

Their frames are three-dimensional, no time factor. In that time, people thought they lived in a three-dimensional world, time was considered to be a completely separate entity. But Einstein realized that a more correct view put time and the three space dimensions on equal footing, setting up a four-dimensional spacetime mode, called Minkowski spacetime.

Gal, Newton's law was shaken passively. Because gravity, like magnetic force, is a static system, **a Braking Force. Even if there is motion, it will stop sooner or later under the continuous and powerful central gravity or magnetic force.** To solve this problem, Einstein constructed the theory of curved space. Space is imagined as a two-dimensional elastic sponge cushion. Heavy celestial bodies bend it, and light celestial bodies revolve around the pit. It's hard to say that it's a stable system, because it doesn't have any power to check centripetal gathering. Can the struggling ships in the ocean maelstrom steer steadily in their courses? Moreover, **Einstein's curved space can only exist in two-dimensional plane, but it is impossible in three-dimensional or four-dimensional space.**

The four dimensional spacetime is easy to understand, it is real. The relativity theory confuses public most is its relativity of time and space, that is, when the object moves at low speed, the cosmos is an inertial frame of reference, obeying the ordinary physical laws; Moving at the speed of light, the cosmos becomes a non-inertial frame of reference, subject to strange laws of physics. Since there are no stationary bodies in nature, any real frame of reference may be considered as an inertial frame of reference only to various degrees of approximation. Observers in different moving frames will measure different relativistic masses of an object as there is no absolute frame of reference with respect to which an object's speed can be measured—Like Newton's default action of distance, Einstein defaulted the observer's space as the absolute reference frame when calculating motions, otherwise, the theory of relativity could not be established. This is also the corpuscular approach, which focuses on isolated objects and isolated motions, ignore the unified field, and therefore requires different reference frames and complex calculations.

Holism, which starts from the whole and links all isolated events with the whole, is an indispensable basic theory of cosmology. **In holism, the cosmos is a whole and the light field is its background, things become simple and clear.**

The field of light was born at the same time as the cosmos and has always enveloped the cosmos. It is the continuum of time, space and energy, sources of all things and provides them with mass & energy. Its space is vast and expanding all the time. It is boundless and concrete for us, because we live in it all the time and sense the presence and changes of it. Its arrow of time never stops, ever goes forward, never backwards; but it is also concrete and measurable to us, for its velocity is definite and constant, viz., C, and its motion is regular and rhythmic. **The light field is our ideal absolute spacetime frame of reference.** From it, we can see several relativistic behaviors in a new light.

- Mass-energy conversion formula: $E=mc^2$
  It can only be true if m is energy particles. This is not true when m is mass particles, atoms, molecules, and objects. For only the former can move at C, and the latter can't unless they are reduced to energy particle.

- Mass increase.
  This is EEEF, or the results of light barrier. Matter does not really increase in motion, which has been discussed before.

- Time dilation.
  This is another way of saying that the inflation of the spherical light field, which results in that **all the waves in the field of light longer and lower in frequency**, and it is inevitable that the velocity of a clock (time) & other things slows down relatively in this space as it flies in sync with the light field. In addition, Newton provided a similar but non-relativistic phenomenon, "Now several astronomers, sent into remote countries to make astronomical observations, have found that **pendulum clocks do accordingly move slower near the equator than in our climates.**" (*The Principia*, 419)

- The twin paradox.
Einstein assumed that one of them was a spectator and fix on earth for the event to occur. In the original sense of

relative, we would **assume both of them were spectators as well as performers**, flying at the same speed at the same time in the light field, their differences in age and appearance would not exist.

- Objects contract.

  As mentioned above, when objects reduce to energy particles in order to reach the speed of light, **their length may not shorten, but disappear.** If this assertion were ignored, and the ruler, the yardstick, the spacecraft, and the twins remained in their original sizes, they too, in the rapidly expanding field of light, would instantly 'shrink' into dots relatively, and disappear into space.

- Back to the past?

  A Superman traveled much faster than light and at 12: 00, he returned to 9: 00 a.m. to retrieve the gold cup he sold at 10: 00. This is absolutely impossible. "What's done cannot be undone." (Shakespeare, *Macbeth*)

  —**The light field can't carry any physical objects, only the images of objects, and the cosmic arrow won't change with the movement of any object.** Superman couldn't find the object (the cup) he was looking for in the light field, but could find the image of the object at most.

  —The light field expands outward at a speed of 300,000 kilometers per second, and all images expand synchronously. Therefore, even if Superman caught up with the image of the Golden Cup, he couldn't see the complete image, because he was trapped in the image of a particle of the cup.

  —When he passed through numerous particles and intended to take a picture of the cup, he found that its image was expanding so fast that it disappeared into space. Even the particle image that had trapped him had now become an infinite vacuum.

- Fly to the future?
  Everything happened, happened, happened in the field of light, and their images were instantly past. Superman saw only images of the past, and the farther he flew, the older images he saw. He could not find an image of the future, because **it had not yet happened.**

- Curved space.
  It is true. Space is curved everywhere, so that our spacecraft cannot fly out of the solar system. But the fact that spacecraft can fly freely in the solar system and get blocked at the edge of it, where the sun's gravity is extremely weak, suggests that **space is warped not by weak gravity but by powerful swirling force. Not only spacecraft, but also a dust can't escape. All objects have to revolve around the sun in their orbits.**

In addition, there is no void. All space is filled with particles and objects that radiate all the time. **The radiation of all bodies is unbalanced, and thus their motion is inherently curved, zigzag, which is the intrinsic cause of the curved space.**

ARGUMENT. Gravity vs Refraction

"**Einstein, Eddington and the 1919 Eclipse**
Peter Coles weighs up three books on the momentous expedition that proved the general theory of relativity.
...Eddington took on the task of proving it. By harnessing a total solar eclipse, he argued that the deflection, or bending, of light by the Sun's gravity could be measured. This was a critical test, because Einstein's theory predicted a deflection precisely twice the value obtained using Isaac Newton's law

of universal gravitation. The needed eclipse came 100 years ago, in 1919. Eddington is now forever associated with two expeditions to view it: from Sobral in northern Brazil, and the island of Príncipe off the coast of West Africa. Those momentous ventures form the kernel of three books commemorating the centenary: *No Shadow of a Doubt* by physicist Daniel Kennefick, *Gravity's Century* by science journalist Ron Cowen, and science historian Matthew Stanley's *Einstein's War*.

Einstein's theory, eight years in the making, sprang from insights he had developed after he published his theory of special relativity in 1905. One of the effects predicted by the new theory was that **light rays passing close to a massive body, such as a star, should be bent by its gravitational field.** This effect had been predicted qualitatively using Newton's theory of gravity. Tantalizingly, Newton himself had written in his 1704 opus *Opticks*: 'Do not Bodies act upon Light at a distance, and by their action bend its Rays…?' But there is no evidence that he calculated the magnitude of the effect (the first full calculation was published by German mathematician Johann Georg von Soldner, in 1804)." (nature.com)

Now let's review a familiar optical principle.
REFRACTION.
"Deflection from a straight path undergone by a light ray or energy wave in passing obliquely from one medium (such as air) into another (such as glass) in which its velocity is different…The root of refraction is seen in the notion that the path of a ray of light or wave of energy is 'broken' when it is deflected or turned. The effects of refraction can be seen in a rainbow, which is formed when light rays passing into (and reflecting out of) water droplets are bent at different angles depending on their color, so that the light separates into bands of color. The amount of refraction depends on the angle and the type of matter; refraction can occur even when passing through different kinds of air. A mirage, such as you might see in the desert or over a patch of asphalt in

the summer, occurs when light passing through warm air meets the very hot air near the surface; reflecting the sky, it often resembles a lake." (merriam-webster.com)

Astronomers tell us that the sun is wrapped in a thick, hot corona, a giant bubble with a diameter of more than 10 million kilometers and a temperature of one million degrees Celsius, and it is no shadow of a doubt that starlight will bend through the corona, making us seen stars behind the sun. It's an explanation much simpler, clearer and more eloquent than that of relativity or gravity.

Two Ends of the Grand Unification: Holism vs Corpuscularianism

"It is an empirical fact that scientific achievements are put just as much, or even more, to destructive as constructive use. The sciences of human behavior and society are no exception. In fact, it is perhaps the greatest danger of the systems of modern totalitarianism that they are so alarmingly up-to-date not only in physical and biological, but also in psychological technology. The methods of mass suggestion, of the release of the instincts of the human beast, of conditioning and thought control are developed to highest efficacy; just because modern totalitarianism is so terrifically scientific, it makes the absolutism of former periods appear a dilettantish and comparatively harmless makeshift. Scientific control of society is no highway to Utopia." (Bertalanffy, 1969, p.52)

All things that exist in the same space affect each other, interact with each other, obey the same laws of nature, must have internal relations, and in some ways may, be unified.

Unity was the universal goal of the ancient Greek philosophers, and it was divided into two schools, holism and corpuscularianism, with corresponding directions up and down, which has been briefly discussed earlier in this book. Now we have to discuss it again, it won't be superfluous.

Scientific research must be inseparable from corpuscularianism, but we cannot go to the end of this road, and eventually need to come back to holism. The reason is that the corpuscularianism is a path that goes down and decomposes. It is the way to eliminate differences step by step and reach an inanimate and rigid end that completely violates the process of the universe and human goal, which is ascending and evolving. Holism is the road to convergence and development.

Take the book market as an example. We can unify its end up, to realize humankind's aim, but we can also unify the end down to the following terms:

1. Words, which eliminate thematic differences between books.
2. Characters or letters, which eliminates the conceptual differences of books.
3. Dots and lines, which eliminate content and form differences of all books.
4. Pixels, which eliminates all differences in all printed materials; books and albums, tapes and CD, junk books and fine books, etc., have achieved a grand unity of publications.

The world scientific path has always been downward. The most typical is physics research, which since the eighteenth century has shifted its focus from atom to particle. The differences between objects, molecules, and atoms disappear. Eventually the bulk of the particle evolves into mathematical points, requiring messenger particles to provide mass & energy.

Another example is biology, where the focus shifts from cells to chromosomes and genes. Soul is gone, life has disappeared, the differences between humankind and other animals and plants are eliminated, and the number and variety of genes of many lower plants and animals are even superior to those of human beings. Others make up for the gap of contact with the upper world by attaching consciousness, intelligence, spirit and soul to corpuscles (genes, cells, atoms, particles, quantum) in a reluctant manner.

More than two centuries of practice on the path of reductionism has proven that it is futile to seek truth in corpuscles; can you expect to find the Coca-Cola formula from particle research? Or find the window system core structure from the careful combination of 0 and 1? Or from genetic makeup reveals N. Paganini's violin skills? At present, the world has come to the end of this narrow path, and its adverse consequences have long gone beyond the scope of scientific research, taking a share of social conflicts and spiritual crises, to the point that some said, **"the finder of a new elementary particle used to be rewarded by a Nobel Prize, but such a discovery now ought to be punished by a $10,000 fine."** (Wilczek, 2008, p.34)

The important reason for choosing holism is that it is closer to the essence of things than corpuscularianism is. As the book, a complete book will best reflect its essence. Down to its words or letters, you can't get it, and down to the pixel, you can't even tell what it is. The purpose of publishing books is to learn and to educate. We must base our task on different but independent and complete books, and unite with the highest end.

Holistic theory is especially necessary for studying the subject of life. Life is a process of independent development, generally consisting of three stages: birth, growth and decay. Life is always in motion, motivated by itself, capable of absorption, digestion, excretion, reproduction, and so on. Life is closely related to the environment as all life originates from the environment and constantly exchanges material, energy and information with the environment. It cannot be separated from the environment at anytime.

These characteristics are most obvious in plants and animals, whose most basic unit of life is the cell, and they determine that life can only exist as a whole. If the whole is destroyed, or broken apart, their characteristics are not fully functioning and can even lead to death.

Nebulae, stars, Earth, oceans, rocks, air, dust, etc., are divisible, and their characteristics of life is incomplete. But like plants and animals, they all contain energy and are generally characterized by motion, vibration and change. They all originate from the environment, and constantly exchange material, energy and information with the environment; that is, all things are interrelated and integrated. Therefore, when we study them, we can't ignore holism.

Solutions: Newton's Five Big Puzzles

"Although I respect all these thinkers and would not wish to make myself disliked by criticizing them, I can give a proof of what I say which I do not think any of them will reject, namely that they have all put forward as principles things of which they did not possess perfect knowledge. For example, there is not one of them, so far as I know, who has not supposed there to be weight in terrestrial bodies. Yet although experience shows us very clearly that the bodies we call 'heavy' descend towards the centre of the earth, we do not for all that have any knowledge of the nature of what is called 'gravity', that is to say, the cause or principle which makes bodies descend in this way, and we must derive such knowledge from some other source. The same can be said of the void and of atoms and of heat and cold, dryness and humidity, salt, sulphur, mercury and all other similar things which some people have proposed as their first principles. Now none of the conclusions deduced from a principle which is not evident can themselves be evident, even though they may be deduced from the principle in an evident manner. It follows that none of the arguments based on such principles have been able to provide their proponents with certain knowledge of anything, and accordingly such arguments have not been able to bring them one step further in their search for wisdom. If they have discovered anything true, it has been solely by means of one of the four methods set out above. Nevertheless, I do not wish to detract in any way from the reputation which any of these philosophers may claim. I am simply obliged to point out, for the consolation of those who have never studied, the following similarity with what happens when we travel: so long as we turn our back on the place we wish to get to, then the longer and faster we walk the further we get from our destination, so that even if we are subsequently set on the right road we cannot reach our goal as quickly as we would have done had we never walked in the wrong direction. The same thing happens if we have bad principles. The more we develop them and the more carefully we work at deducing various consequences from them in our belief that we are philosophizing well, the further we move from knowledge of the truth and from wisdom. The conclusion that must be drawn from this is that among those who have studied whatever has been called philosophy up till now, those who have learnt the least are the most capable of learning true philosophy." (Descartes, 1985, p182-183)

Newton advocated experimental philosophy, and the corresponding method was to provide the proper description of the relevant natural phenomena, and then find the most general principles that account for them. By this way, he founded the law of universal gravitation, but this method failed to make him know the essence of gravitation and also caused many puzzles, mainly the following.

A. Action at a distance

"That gravity should be innate, inherent, and essential to matter, so that one body may act upon another at a distance through a vacuum, without the mediation of anything else, by and through which their action and force may be conveyed from one to another, **is to me so great an absurdity that I believe no man who has in philosophical matters a competent faculty of thinking can ever fall into it.** Gravity must be caused by an agent acting constantly according to certain laws; but whether this agent be material or immaterial, I have left open to the consideration of my readers."

B. Transverse Motion of Planets

"the compound of this attraction & projection would according to my notion cause a circular revolution of the earth about the Sun. But the transverse impulse must be of a just quantity, for if it be too big or too little it will cause the earth to move in some other line. Secondly **I do not know any power in nature which could cause this transverse motion without the divine arm."**

C. Stability of the Solar System

Newton realized that gravitational systems are precise and fragile, so that mere chance, or the ordinary physical interactions of the planetary bodies, would have produced an unstable planetary system, one in which the planets would eventually either be too strongly attracted to the sun, falling into it, or be too weakly attracted, flying off into space.

"And much harder it is to suppose that all the particles in an infinite space should be so accurately poised one among another as to stand still in a perfect equilibrium. for I reccon this as hard as to make not one needle only but an infinite number of them (so many as there are particles in an infinite space) stand accurately poised upon their points—**stand upright on its point upon a looking glass."**

D. The formation of the solar system

"As to your first Query, it seems to me that if the matter of our Sun & Planets & all the matter in the universe was evenly scattered throughout all the heavens, & every particle had an innate gravity towards all the rest & the whole space throughout which this matter was scattered, was but finite: **the matter on the outside of this space would by its gravity tend towards all the matter on the inside & by consequence fall down to the middle of the whole space & there compose one great spherical mass** But if the matter was evenly diffused through an infinite space, it would never convene into one mass but some of it convene into one mass & some into another so as to make an infinite number of great masses scattered at great distances from one another throughout all that infinite space. And thus might the Sun and ffixt stars be formed supposing the matter were of a lucid nature. But... **Why there is one body in our Systeme qualified to give light & heat to all the rest I know no reason but because the author of the Systeme thought it convenient,** & why there is but one body of this kind I know no reason but because one was sufficient to warm & enlighten all the rest."

These four puzzles were expressed in Newton's correspondence with Pastor Richard Bentley from January, 1692 to February, 1693.
(https://www.sophiararebooks.com/pages/books/3537/sir-isaac-newton/four-letters-from- sir-isaac-newton-to-doctor-bentley-containing-some-arguments-in-proof-of-a-deity)
He admitted that he could not find good explanations for these and handed them over to God. In addition, I think there was a fifth puzzle.

F. Antigravity

There are **three kinds of antigravity.** The first is reaction force, which is determined by Newton's third law. Newton could not ignore this, just rarely use it. The second is the centrifugal force. This word appeared 30 times in *The Principia* (in contrast, centripetal force is 213 times). Newton calculated that it was 2 million times stronger than the combined gravity of the sun and the moon (468), it raised water levels at the equator by 85472 Paris feet, which is about 21km, over those at the poles (465). He plainly declared that centripetal force was 'a gravitating force' (400), but never plainly declared that centrifugal force was an antigravity. The third is radiation proposed in this book, which Newton might not have thought of. In *The Principia*, **the word 'radiation' is used only two times** when it comes to the tails of comets. (*The Principia*, 510) Gravity is very weak, uncertain, and negligible, whereas **radiation is very**

powerful and certain. This alone is enough to overturn the law of gravity.

If we would like to think the following question, we might understand why Newton shunned the fact of radiation.

According to Newton's law, stars and planets were huge gravitational bodies, and everything on the earth's surface, including mountains and seas, was firmly attracted. Newton asserted that light was matter, corpuscles, and the mass of the sun was 330,000 times greater than the mass of the Earth, **so why could light and other corpuscles escape from the sun all the time, in large quantity, and not be attracted?** Radiation was the opposite of gravity, and that might be the answer.

These five questions boil down to one: what is gravity?

In *The Principia*, Newton's answer was not so clearcut. At first, he emphasized centripetal force, which consists of three forces. Among them, the second is "magnetism, by which iron tends to the loadstone." The first is "gravity, by which bodies tend to the centre of the earth." The third has no name, "and that force, what ever it is, by which the planets are perpetually drawn aside from the rectilinear motions, which otherwise they would pursue, and made to revolve in curvilinear orbits." (*The Principia*, 3)

Later, the first force and the third force become one.

"PROPOSITION V. THEOREM V. That the circumjovial planets gravitate towards Jupiter; the circumsaturnal towards Saturn; the circumsolar towards the sun; and by the forces of their gravity are drawn off from rectilinear motions, and retained in curvilinear orbits."(Ibid.,399)

"SCHOLIUM. The force which retains the celestial bodies in their orbits has been hitherto called **centripetal force;** but it being now made plain that it can be no other than a gravitating force, **we shall hereafter call it gravity.**" (Ibid.,400)

Even so, The puzzle of what gravity is remains unsolved. (530) Now let's go back to the beginning of *The Principia*, and focus on two forces: vis insita, or innate force of matter and impressed force. The characteristic of the latter force is,

"This force consists in the action only; and remains no longer in the body, when the action is over. For a body maintains every new state it acquires, by its vis inertiæ only. Impressed forces are of different origins as from percussion, from pressure, from centripetal force." (*The Principia*, 3)

Centripetal force or gravitation is definitely not 'vis insita', which Newton made clear again in his correspondence with Richard Bentley. "You sometimes speak of gravity as essential & inherent to matter: pray do not ascribe that notion to me, for the cause of gravity is what I do not pretend to know, & therefore would take more time to consider of it."

Newton also denied action at a distance. In the solar system, there is a vast space between the sun and the planets. Gravity is an impressed force at a distance to every celestial body. How can it happen? **Newton failed to answer it. This may be the cause why Newton hereafter discontinued physical research and turned to alchemy and theology, and Later, accepted the appointment as Warden of the Mint with pleasure.**

Having said that, we can be sure that Newton never ceased to think and explore the puzzle after that. If he found the answer, he would not keep it to himself. It is safe to say that he never found the answer, that no one else has provided the answer, and that gravity is still unproven and only a hypothesis.

It is true that Newton provided a great deal of natural phenomena in his book and made precise calculations; For more than 300 years now, Newton's followers (I was one of them) have provided a wealth of new discoveries and evidence in support of Newton's laws, including the discoveries of Neptune & Pluto, but none of this proves the existence of gravity.

I agree with Newton on this point: Gravity is not essential & inherent to matter. It consists in the action only; and remains no longer in the body, when the action is over. I would add that **they interact only in the case of direct contact & resonance.**

Also, weight is not innate force of matter, we have to put objects on the platform scale to produce weight, which is the result of the two objects interacted. By the same token, the celestial bodies have to contact each other before gravity can be occurred. Contact is not radiation, which is antigravity; nor approach, as described in *The Principia*, but the zero distance connection between objects. Take the earth and the moon as examples, the moon must enter the earth's atmosphere at least, and the mutual attraction will take place. That is, **there is no gravity (and any force) between objects in isolated state, the universal attracting force between everything is non-existence.**

This book already provides answers to the puzzles, and now I gives the following hints:

A. The **light field can eliminate action at a distance** and transmit gravitational waves, if any.

B. Planets burst out from the sun to obtain **angular momentum** for spinning and rotating around the sun under the constraint of shells.

C. **Vortex motion and shells keep the universe stable and harmonious.**

D. **The cosmic force, vortex motion and black hole constantly making new celestial bodies and systems.**

E. Strong and universal antigravity: radiant force and tension. Since the cosmic background radiation is close to absolute zero, all objects have spontaneous radiation; Since atoms have their own independent space, all objects will produce equal counter pressure when pressed, that is, tension. **Radiation force and tension are both antigravity, and the sum of them is greater than gravity.**

Paraphrase Newton's Forces

"But I, who am well aware that no judgment can be passed on uncommon or remarkable things, much less anything new brought to light, unless the causes of common things, and the causes of those causes, be first duly examined and found out, am of necessity compelled to admit the commonest things into my history. Nay, in my judgment **philosophy has been hindered by nothing more than this, that things of familiar and frequent occurrence do not arrest and detain the thoughts of men, but are received in passing without any inquiry into their causes;** insomuch that information concerning things which are not known is not oftener wanted than attention concerning things which are." (Bacon, 2018, BOOK I,119)

Particle physicists excel at using mathematical techniques to build mathematical models from phenomena and experiments to solve practical problems, and they have achieved much in this way. Surprisingly, they are often ignorant or misunderstand the causes of phenomena.

Newton, for example, invented the law of gravitation and it was widely used, but he never explained its cause clearly. In fact, he admitted not knowing. To avoid embarrassment, he was trying to substitute **attraction** for gravity.

Using the computing function of WORD, I found that the term of 'gravit' (include gravity, gravitate, gravitates, gravitating, gravitation) **appeared 457 times** in the text of *The Principia,* And 'attract' (include attracts, attracted, attracting, attractive, attraction) **appeared 521 times,** and Newton didn't consider they were synonyms. The reason why he did this can be understood from the following passage.

"SCHOLIUM.

These Propositions naturally lead us to the analogy there is between centripetal forces, and the central bodies to which those forces used to be directed; for it is reasonable to suppose that forces which are directed to bodies should depend upon the nature and quantity of those bodies, as we see they do in magnetical experiments.

And when such cases occur, we are to compute the attractions of the bodies by assigning to each of their particles its proper force, and then collecting the sum of them all. I here use the word **attraction** in general for any endeavour, of what kind soever, made by bodies to approach to each other; whether that endeavour arise from the action of the bodies themselves, as tending mutually to or agitating each other by spirits emitted; or whether it arises from the action of the aether or of the air, or of any medium whatsoever, whether corporeal or incorporeal, any how impelling bodies placed therein towards each other. In the same general sense I use the word **impulse,** not defining in this treatise the species or physical qualities of forces, but investigating the quantities and mathematical proportions of them; as I observed before in the Definitions." (188)

Like gravity, attraction is not 'vis insita, or innate force of matter', but a force of a variety of external 'impulse'. "**a centripetal force...I likewise call attractions and impulses, in the same sense, accelerative, and motive; and use the words attraction, impulse or propensity of any sort towards a centre, promiscuously, and indifferently, one for another;** considering those forces not physically, but mathematically." (6)

Centripetal force is generally regarding as gravity, as Newton ascertained "centripetal force... we shall hereafter call it gravity." (400) So gravity is the dominant force in all celestial motions. Whereas there is no gravity in the above quote (6), attraction or impulse are regarded as the real centripetal force. It's not very convincing, but a little clearer than gravity for explaining the cause of centripetal force.

From our common sense, **every force consumes energy, or needs external force (Newton: impressed force) to push.** The energy or external force of attractive force should be gravity. But there is no gravity among the external causes in the above SCHOLIUM. So we can only understand the relationship of the three forces in this way: in physics, centripetal force is gravity; and mathematically, it is attractive force.

In my opinion, **centripetal force, gravity and attractive force are synonyms, all of which are virtual forces and do not exist in reality. Because their causes are inexplicit, actually void, and they all have the paradox of action at a distance.**

Furthermore, energy is the source of force and is inherent in matter; Force is the application of energy, a phenomenon that occurs only when bodies interact with each other.

In physics, there are only one energy & one force: Cosmic Energy, viz., vibration force and its flight state, radiation, which is the eternal forces in everything, Newton's real 'vis insita, or innate force of matter.'

Force does not exist independently, and it always coexists with matter and energy, "every force is conditioned and requires something else than itself for its subsistence." (Hegel, *Shorter Logic*, §136) Energy pushes objects to move, objects apply forces, which is converted from energy, to interact with each other, and their particles vibration is the basic eternal energy or force of objects—When an object is in an isolated state, its vibration appears as energy; When multiple objects interact with each other, their vibrations become forces.

A common phenomenon we see in our daily life is that energetic bodies such as a coil of spring, a spinning top (a gyro), a lit firecracker, their motions are not rectilinear line, but curve. For the same reason, the motion paths of all stars (plasma) and galaxies is curved. Their radiation disequilibrium is the main cause of their curved motion.

Another thing energetic bodies have in common is that they rely on releasing their own energy to move, so they are radiators, centrifugal motion bodies. The energy of spring, top, and firecracker is artificial and soon exhausted, and their movements cease at the same time. The energy of plasma is eternal, and the stars can glow, spin & move in their orbits permanently. It is also a centrifugal force, a divergent force, that the shell reverses its direction of motion, let them live a stable and long life.

The principles of vibration and vortex motion are clear and self-evident, and it can be determined by direct observation that they maintain the stability and harmony of the universe. Most astronomical phenomena can be satisfactorily explained by them.

Resonance is a more powerful and accidental force, and it is responsible for most contingencies.

From this discussion we find another cardinal difference between deduction and reduction. The former focuses on essence & causes, the latter on phenomena & effects.

The main subject of classical physics is forces, isolated forces, that have nothing to do with energy. They have been calculated very finely and precisely, and have been widely used in practice with remarkable results, especially in electricity and nuclear forces. However, most of them are phenomena without causes, know how, don't know why, so there are many paradoxes, the forces without causes, which hinder further development.

Here, I think it is useful for us to revisit the saying of American essayist and poet Ralph Waldo Emerson.

"Cause and effect, means and ends, seed and fruit cannot be severed; for the effect already blooms in the cause, the end preexists in the means, the fruit in the seed." (GIGA Quotes)

The Road to Ultimate Theory

"string theory, in particle physics, a theory that attempts to merge quantum mechanics with Albert Einstein's general theory of relativity. The name string theory comes from the modeling of subatomic particles as tiny one-dimensional 'stringlike' entities rather than the more conventional approach in which they are modeled as zero-dimensional point particles. **The theory envisions that a string undergoing a particular mode of vibration corresponds to a particle with definite properties such as mass and charge. In the 1980s, physicists realized that string theory had the potential to incorporate all four of nature's forces—gravity, electromagnetism, strong force, and weak force—and all types of matter in a single quantum mechanical framework, suggesting that it might be the long-sought unified field theory.** While string theory is still a vibrant area of research that is undergoing rapid development, it remains primarily a mathematical construct because it has yet to make contact with experimental observations." (Encyclopedia Britannica)

"Albert Einstein produced his General Theory of Relativity. It describes gravity as the curvature of 4 dimensional spacetime by massive objects. It has been proven to describe the way planets orbit the Sun and other large scale interactions.

Quantum mechanics describes the interactions of particles on a very small scale. It describes electromagnetism, radioactivity and how atoms are assembled.

There is a problem. The two theories are totally incompatible. Black holes and the Big Bang require the two theories to work together, which they don't. Also, there are many unanswered questions, like why is a muon 200 times more massive than an electron.

So, a Theory Of Everything is required to unify the two theories. String theory is such a theory. The actual name is **super string theory. The idea is that fundamental particles like electrons and quarks are actually a vibrating one dimension string in a multi-dimensional space.** There are actually several string theories.

The various string theories are being united by a theory called M theory. It requires strings and multi-dimension structures called **branes (for membrane) which exist in an 11 dimensional universe.**

One problem with string theory is that the mathematics is very very hard. Even approximations are very difficult to solve.

We definitely need **new physics to unify General Relativity and Quantum mechanics.** Maybe string theory will be the answer." (**'String theory is a candidate Theory Of Everything'**, by Phillip E. May 17, 2017. Socratic.org)

Vibration and its resonance is cardinal factor of grand unity of matter & energy, which is why I can't ignore string theory. But obviously, string theory can't be TOE. Firstly, it can't explain the original power of vibration & resonance of all things. In fact, there is no theory other than CFT that can answer this question. Secondly, it lacks the notions of radiation & field. Their relation is that **vibration of particles is through radiation to form a field that induces resonance.**

The four forces are very different not only in strength but also in direction. Electricity and weak force are outward, while magnetism, gravity, and strong force are inward. They can never be unified. The ultimate theory aimed at the unification of the four forces can only be a dream.

The vortex force is real, it is the most universal and most powerful force in the universe. All four forces appear insignificant before it, and all collapse in it.

Not only force but also matter, large or small, can be crushed by the force of the vortex and fall into a black

hole, becoming energy at last. But the destructive force of the vortex is also the force of construction. It causes the black hole to light up and create new stars. This would also allow stars to explode, creating small stars, planets and moons, that might create the conditions for life.

Vortex force comes from the combined forces of the particles, all of which are photons. Light is the cosmic force, an eternal vibrating & radiating Ether, which constitutes the mass & energy of all things. Everything comes from light and is destroyed by light. The cosmic process is the movement and evolution of light field. This is the true ultimate theory.

An inquisitive person may ask, where does the light field come from?

The answer provided by CFC is as follows: **the light field is the most primitive existence and has no source, it is the background field of the space. In other words, the universe, the space, is not void, it itself is the light field, the light ether.**

The idea comes from ancient Greek philosophers. They believed that nothing was created out of nothing, and that nothing was destroyed so as to become nothing. Therefore, everything has a source, but the most primitive thing has no source[1], it is the arche (author) of everything. "Philosophers talk about 'first cause uncaused,' or 'first mover unmoved.'"

---

[1] "It is plain then that they all in one way or another identify the contraries with the principles. And with good reason. **For first principles must not be derived from one another nor from anything else, while everything has to be derived from them.** But these conditions are fulfilled by the primary contraries, which are not derived from anything because they are primary, nor from each other because they are contraries." (Aristotle, *Physics*, BOOK I, Part 5)

"REGARDING this kind of substance, what we have said must be taken as sufficient. All philosophers make the first principles contraries: as in natural things, so also in the case of unchangeable substances. But since **there cannot be anything prior to the first principle of all things, the principle cannot be the principle and yet be an attribute of something else.**" (*Metaphysics*, p.157)

(Durant, 1991, p.125) They came up with a variety of theories of the cosmic arche, and Democritus' atomism emerged as the dominant one.

A primitive thing ought to be single and indivisible, and an atom was considered such thing at that time. However, with the advance of science, people find that atoms are divisible, and of more than 100 species, its status was challenged by the singularity theory.

The primitive universe has only one singularity, which is the source of everything; I call it matter-based theory (MBT). Light field theory is the energy-based theory (EBT), energy is the source of all things, and energy-mass interaction is the basic motion of the universe. Both theories, though radically different in nature, are the original theories of the universe and their objects have no source, which is what they have in common.

Although singularity theory is a novel original theory, and singular point is also the ultimate goal of vortex motion, it is incompatible with space conditions. Singularity can neither contain space nor create space, and space is innate and indispensable to all existence. The light field is completely compatible with space without any contradiction and thus, it is the origin of the universe.

It's a cycle of light: Light field ➤ Free light (photon) ➤Semi-confined light (ion, star, electricity) ➤Completely confined light (atom)➤Multiple confined light (molecule, stuff) ➤Free light (radiation, burning, explosion) ➤ Light field.

The motion of the universe can be shortened to a phrase: the endless cyclical evolution of light, the true GUT of the universe. We can see it as a vindication of the third great achievement of Ancient Greek speculative philosophy, Heraclitus' theory of **Central Fire.**

"Herakleitos and Hippasos say that **the first principle of all things is fire; for they say that all things arise from fire**

and they all end by becoming fire. As this is quenched all things come into the order of the universe; for first the dense part of it contracting into itself becomes earth, then the earth becoming relaxed by fire is rendered water in its nature, then it is sublimated and becomes air; and again the universe and all bodies are consumed by fire in the conflagration. **Fire then is the first principle because all things arise from this, and the final principle because all things are resolved into this.**" (Heraclitus, *Fragments.* Translated by J. Burnet, A. Fairbanks, and K. Freeman. wordpress.com.)

First principle has no source, is absolute and eternal truth. The Greeks bravely search for first principles and the highest knowledge of the universe, while modern people are rapt in searching for truth among corpuscles, which is why we can hardly surpass them.

Ancient Greeks believed that Muses dwell on aloft. The higher our souls soar, the closer we get to them. This is the shocking secret of their creativity.

Wisdom Is the Author of Truth and Knowledge

It is astonishing with how little wisdom mankind can be governed, when that little wisdom is its own.
William Ralph Inge (GIGA Quotes)

Francis Bacon said **"knowledge is power."**

Plato said **"wisdom is the author of truth and knowledge."**

These are two different worldviews. In reality, the world has been following Bacon's direction. The most important achievement is nuclear weapons, which are the result of cutting-edge knowledge and are the greatest power on Earth. Whoever possesses the most advanced and the largest number of nuclear weapons is the most powerful and secure.

However, it is widely recognized by contemporary people that nuclear weapons, though powerful, cannot guarantee the safety of anyone and are more likely to destroy all together. Even if a few survive, their fate will be worse than death.

The only right direction for the world to move in is the path of wisdom. **Knowledge is neither good nor evil and has no soul, but anybody with a soul can use it to do good or evil.** A man with all the knowledge of the British Museum may be a Holmes or a Moriarty. The essence of knowledge is the means, not the end. What human beings need to cherish most is wisdom. If they abandon wisdom, they will inevitably make mistakes. The world has been on the wrong path. The scramble to develop and upgrade destructive weapons is a prominent example in support of this assertion.

People are intelligent creatures with initiative. The road of knowledge is not the only road in the world. The world can choose a better road, and human beings can lead a better life, just entailing wisdom.

What is wisdom? Although wisdom is a familiar term, I dare say that not many people have a profound understanding of it yet. I have reduced it to three abilities: discrimination, choice, and creativity. The Bible tells us that wisdom is the ability to distinguish good from evil. To broaden that definition, it is the ability to distinguish beauty from ugliness, good from evil, and truth from falsehood. However, this is only a kind of cognitive ability, is not enough. It must be combined with practical ability, that is, **the ability to love beauty and hate ugliness, to promote good and discard evil, to long for truth and stay away from falsehood.** This is the ability of choice. If there is no choice, we will devote ourselves to artistic creation and create beauty & goodness. That's what **Aristotle calls practical wisdom, 'phronesis' in Greek. This is the most important thing for both the individual and the world.**

Why are the Greeks able to independently create philosophy, science and mathematics? Most important of all, they pursue first principles in unison. Although limited by means of observation and social context, their views were not unified, but it is the universal wisdom approach. Later generations are following their direction, just lack of discernment. Most choose Democritus' atomism and develop it into particle theory and quantum theory. Modern astronomical observations lead us to believe that Leucippe is right about the vortex. One step further, it is not hard to find that the force that drives the vortex is the cosmic force, and Heraclitus' central fire (light) is the first principle.

"For a fact it's nothing except in relation to desire; it is not complete except in relation to a purpose and a whole. Science without philosophy, facts without perspective and valuation, cannot save us from havoc and despair. Science gives us knowledge, but only philosophy can give us wisdom." (W.Durant) We must study Greek philosophy afresh if we want to achieve more in science and in other things.

The Greeks regarded wisdom as the second of the five virtues, the first being piety. If we replace knowledge with wisdom, **when people generally possess these three abilities, beautiful, good, and true things can emerge in large numbers, and ugly, evil, and false things gradually disappear without severe punishment or harsh administrative regulations, and the world will move toward harmony and beauty.**

The Task of Science: Discovery or Design?

To raise new questions, new possibilities, to regard old problems from a new angle, requires creative imagination and marks real advance in science.
Albert Einstein
(https://www.pinterest.nz/pin/708613941079708291/)

Nature is an objective reality. The mission of science is to Through the observation of natural phenomena, crack the mystery of nature, finally discover the truth of nature. This is the consensus of most scientists. Observatories, laboratories, a number of publications and research institutes named after discoveries, were established for this purpose.

Mathematics is a handy tool for scientific research, and the mathematical model is an indispensable means for scientific discovery. It and other models are not natural but are designed subjectively, and the task of science is modified accordingly: to build mathematical models consistent with natural phenomena.

The mathematical model is designed. In reality, what we do is to design before we practice, such as building houses, bridges, roads, and even countries. This is also true for scientific research. Hence Hawking's idea expressed in *The Grand Design*: **there is only 'model-dependent realism', and there is no 'model-independent realism'.** (2011) Whatever your model is, the world reflects it, just like a house is built based on a design. Accordingly, there is no natural truth. Both geocentric and heliocentric theories have their own models to support them.

Hawking's idea comes from the long scientific practice of humankind. People have always believed that nature is model-independent reality, but in fact it is a complete model-dependent reality, starting with ancient Greeks. The Greek philosophers' theories of water, fire, the four elements, and the numbers, are models of the universe.

Later, geocentrism and heliocentrism, up until the Big Bang and the expansion of the universe, were models of the universe. Abandoning these models left us with only Lao-tzu's Taoism and creationism.

Hawking's freedom to create a model of the universe for all people has become the basis for my establishment of force cosmology, which is a brand-new model of the universe. Scientific discovery requires knowledge and observational tools, and the establishment of models cannot happen without these two conditions. But people who build models do not necessarily have to observe and do experiments themselves. They can borrow other people's research results. Ordinary people can also undertake this task.

With this idea in mind, I hope that readers will not focus on my qualifications but on whether my model conforms to objective reality, as this statement deviates slightly from Hawking's original intention.

I call the cosmic force theory (CFT) and the cosmic force vortex theory (CFVT), absolute truth. **I don't expect many people to agree with me, I'll be content to let CFT & CFVT have been listed in lines of science and philosophy.** The last paragraph of Stephen Hawking's *A Brief History of Time* expresses my wish.

"However, if we do discover a complete theory, it should in time be understandable in broad principle by everyone, not just a few scientists. Then we shall all, philosophers, scientists, and just ordinary people, be able to take part in the discussion of the question of why it is that we and the universe exist. If we find the answer to that, it would be the ultimate triumph of human reason—for then we would know the mind of God." (2016)

Big Bang Model vs Cosmic Force Model

In accordance with Hawking's statement in *The Grand Design*, I do not comment on modes, but only list the main arguments of the Big Bang mode and the cosmic force mode for the reader's judgment. The Following are views of the big bang mode, code named B (Big Bang), compared with my views, code named F (force).

- Cosmic entities

B: Isolated corpuscles; the total is 10^{80}.

F: Unified field, light field, and the physical world; the total is 2.

- Matter and particles

B: Matter is particles. Particles are only mathematical points. Everything about them is foreign, including mass (which creates gravity) and energy, which are provided by messenger particles.

F: Matter is atoms. Particles are the cosmic force, and their aggregation constitutes atoms. **Atoms are trapped particles that make up the mass of bodies and their material energy.** Free particles are energy, they combine into a unified energy field.

- Antiparticle and antimatter

B: At the moment of the Big Bang, each particle had an antiparticle, and the two annihilated each other. Since there are a few more particles than antiparticles, everything in the universe came into being. Positrons, negative protons, and antineutrons have been found to form antimatter, and **there may be antiobjects, antihumans, or even an antiuniverse,** in space that have not yet been observed. But an antimatter bomb has already been established theoretically, and it's more powerful than nuclear weapons and therefore more capable of destroying the world.

F: Light diverges into space in a sphere; particles and **antiparticles are just spherically symmetric phenomena. Antiparticles are everywhere, but without any antimatter,** and antimatter bombs are fantasized apocalyptic annihilation tools that will never be produced.

- Wave-particle duality of light

B: Photons have both wave and particle natures, sometimes showing waves and sometimes particles. Bosons are characterized by volatility; fermions are characterized by their particle nature.

F: When photons travel at the speed of light, they are not matter but energy, showing volatility. When the same photons travel below the speed of light, they are matter and not energy and act as particles.

- Mass-energy conversion

B: When all matter moves at the speed of light, it is converted into energy according to the formula $E=mc^2$. At the same time, their mass becomes infinite. (Note: this is a big paradox.)

F: **When all particles move at the speed of light, they are converted into cosmic energy, and merge into the light field.**

- Energy-mass conversion

B: Energy can be converted into matter, but this phenomenon is rare, **"about one particle per cubic kilometre per year."** (*A Brief History of Time*, p.56)

F: This phenomenon is common. Particles are converted into matter when their speed drops below the speed of light in their interactions with matter.

The latest interpretation. **there are only one energy & one force in cosmos: vibration, which forms the**

particle vibration spectrum. When a particle accelerates to the left of the spectrum, it moves towards a mass-energy conversion; To the right, to energy-mass conversion.

- The birth of the universe
B: The universe was created from a singular explosion. This explosion caused the universe to expand, and the expansion movement continues to this day, and has been accelerating the whole time.
F: The universe is generated from the ignition of **a giant black hole,** which caused a brief expansion of the universe. After the vortex movement formed and came to be dominant, expansion movement turned into contraction movement, which continues to this day.

- The general motion of celestial bodies
B: Celestial bodies are radiating outward in a straight line from the centre of the Big Bang, while rapidly increasing their distance from each other.
F: Celestial bodies generally carry out a swirling centripetal curvilinear motion, which gradually shortens the distance between them. However, frequent explosions in the process frequently slow down the rate of contraction.

- Dynamics of the overall motion and form of celestial bodies
B: **Gravity.** It is directly proportional to the mass and inversely proportional to the square of the distance. The main form of motion is mutual attraction of straight lines.
F: **Cosmic force, LIGHT.** That is, the accumulation force of particles is in direct proportion to mass. The main form of motion is curvilinear nested convolutions, that is vortex motion.

- Types and properties of forces
- B: There are four kinds: **strong force, weak force, electromagnetic force & gravitational force.** They are messenger particles hidden in matter that are released when atoms split or decay, and displayed when objects interact — virtual photons are often required as mediums. The original source of force is the God particle, and the energy comes from the first push.
- F: There are **two kinds: vortex force and interaction force.** Vortex forces are dominant, forcing objects come into contact and interact with each other, mainly in the form of aggregation & resonance. The original source of force is **LIGHT,** and all motion of matter is the transformation form of the speed of light, which is the cosmic energy.

- Action at a distance and gravity
- B: **Action at a distance is negated in theory and applied in practice.** In the quantum mechanical way of looking at the gravitational field, the force between two mass particles is pictured as being carried by a particle of spin 2, called the graviton. This has no mass of its own, so the force that it carries is long-range. The gravitational force between the Sun and the Earth is ascribed to the exchange of gravitons between the particles that make up these two bodies. (Ibid., 79)
- F: **Action at a distance is negated theoretically & practically.** Gravity is the polymerization force generated by the contact between molecules and atoms of matter under the action of pressure, which is proportional to the pressure and the number of molecules and atoms in contact with each other.

- Generation of electromagnetic force
- B: The electromagnetic attraction is pictured as being caused by the exchange of large numbers of virtual

massless particles of spin 1, called photons. Again, the photons that are exchanged are virtual particles, (Ibid., 80)

F: There are magnetism & electricity in nature. Magnetism is ordered corpuscular resonant vibration force; electricity is ionic resonant vibration force. They are connected and induced with each other, but **the unified electromagnetic force & field is untrue.**

- Characteristics and types of fields

B: A field is a set of physical properties that exist at every point of space, **each quantum is a spread-out unit of field with its own discrete energy.** There are five force fields: gravity, electromagnetic forces, strong, weak nuclear forces, and Higgs field. Two matter fields: lepton and baryon fields.

F: Cosmic field is formed by **LIGHT.** It's pure energy, invisible, massless, and enveloping the universe. It spreads out as waves at the speed of light. It interacts with all matter, polymerizes with them, results in various **material fields,** such as the atmosphere, the ocean, the land and so on. There are also many kinds of micro- fields inside matter, such as the electron field and the quark field. The field of light creates space without boundary; Matter fields are energy trapped in matter and have each boundary.

- Formation of stars

B: Stars are formed by nebula accumulation under the action of gravity.

F: Stars are formed by swirling nebulae under the action of vortex force that go through the black hole stage.

- The causes and results of stars luminescence

B: "They would remain stable in this state for a long time as stars like our sun, **burning hydrogen into helium and radiating the resulting energy as heat and**

light. More massive stars would need to be hotter to balance their stronger gravitational attraction, making the nuclear fusion reactions proceed so much more rapidly that they would use up their hydrogen in as little as a hundred million years... it seems likely that the central regions of the star would collapse to a very dense state, such as a neutron star or black hole." (Hawking, 2016, p.135-136)

F: **Stars glow when their gaseous ions are compressed into liquid and solid plasma, no nuclear fusion occurs,** the energy loss is minimal, and they do not age and die.

- The essence of the spiral arm

B: It is not a real material arm, **it is just a density wave pattern** formed by celestial bodies in the process of orbiting the galactic centre.

F: **It is a real material arm consisting of celestial bodies.** It transports bodies to the central black hole like the conveyor belt of a blast furnace. In addition, the spiral arm itself causes a helical centripetal precession.

- The nature of black holes

B: **They are cryogenic lifeless bodies formed by the collapse of aging giant stars.** "A region of space-time, from which nothing, not even light, can escape, because gravity is so strong." (Hawking, 2016, p.236) A fallacy: "They all confirm that a black hole ought to emit particles and radiation as if it were a hot body with a temperature that depends only on the black hole's mass: the higher the mass, the lower the temperature." (Ibid, 2016, p.119-120)

F: A black hole is the state in which matter at the centre of a vortex is accelerated to the speed of light. **It is the precursor of a new celestial body, full of vitality.** It is a transitional body and no longer exists after the birth of the new body, but big black holes last forever. It is the mass–energy converter and energy–mass converter of the universe; it is **the celestial furnace.**

- The nature of quasars
B: Quasars are strange objects with extremely high luminosity that are aging and are far away from us. A quasar has the characteristics of small size, high energy, large redshift value, and instability. Most of them were born in the early universe and are now on the edge of it. Its escape velocity is extremely high, and even many times faster than the speed of light.
F: A quasar is a phenomenon in which the spiral arm coils objects on it to the tip of the spiral arm, and the edge of a black hole, and rips them into pieces and strips by the force of the black hole, resulting in a astonishing dazzling band of light that form **a bright border of a black hole,** and quickly fall into it. Like black holes, quasars are everywhere in space, but their life span is generally short, and their shape and brightness change rapidly.

- The concept of neutron stars
B: Neutron stars are formed by the collapse of an old superstar. They are characterized by low temperature, high spin-speed, high density, and strong pulse.
F: **The neutron star does not exist in reality and cannot be established in theory.** Almost all the pulse signals observed at present originate from quasars and black holes rapid spin.

- The protagonists of the two theories
B: **Corpuscles** (atoms, protons, electrons, quarks, etc.) and their shadows (quantum) are independent, positive and negative, with a large number of transmission force and interaction messenger

particles — virtual photons. Scientists intend to find the truth of the universe from their craziest, most vehement state, driving them to fly and collide at a high speed. The universe is the particle world or quantum world, the task of scientists is to discover truth of the universe through observation, calculation and experiment, and build up the unified cosmic model on the basis of corpuscles.

F: **Man is the observer.** Apply human common sense, reason and wisdom to observe the universe, regard everything in the universe as an interrelated and mutually influential whole, and determine the supreme position and mission of human beings in the universe. The explorers believe that we are the universe, man is the measure of all things, and the truth of the universe lies in our hands.

A Strong Pillar of CFT ➤ Systems Science

Order and system are nobler things than power.
John Ruskin

Atomism, particle theory and quantum theory all belong to reductionism, and the 20th century is its most brilliant era. However, in the upbeat note, a few scholars are aware of its limitations and crisis, and strive to find a way out. Systems theory came into being in this context, and the acknowledged founder is an Austrian-born biologist **Ludwig von Bertalanffy.**

REDUCTIVE SYSTEMS THEORY

"**Modern science is characterized by its ever-increasing specialization, necessitated by the enormous amount of data, the complexity of techniques and of theoretical structures within every field.** Thus science is split into innumerable disciplines continually generating new subdisciplines. In consequence, the physicist, the biologist, the psychologist and the social scientist are, so to speak, encapsulated in their private universes, and it is difficult to get word from one cocoon to the other." (Bertalanffy, 1969, p30)

"**General system theory, therefore, is a general science of 'wholeness' which up till now was considered a vague, hazy, and semimetaphysical concept.** In elaborate form it would be a logicomathematical discipline, in itself purely formal but applicable to the various empirical sciences. For sciences concerned with 'organized wholes,' it would be of similar significance to that which probability theory has for sciences concerned with 'chance events'; the latter, too, is a formal mathematical discipline which can be applied to most diverse fields, such as thermodynamics, biological and medical experimentation, genetics, life insurance statistics, etc." (Ibid., p.37)

"My first example is that of closed and open systems. Conventional physics deals only with closed systems, i.e., systems which are considered to be isolated from their environment. Thus, physical chemistry tells us about the reactions, their rates, and the chemical equilibria eventually established in a closed vessel where a number of reactants is brought together. Thermodynamics expressly declares that its laws apply only to closed systems. In particular, the second principle of thermodynamics states that, **in a closed system, a certain quantity, called entropy, must increase to a maximum, and eventually the process comes to a stop at a state of equilibrium**...However, we find systems which by their very nature and definition are not closed systems. **Every living organism is essentially an open system. It maintains itself in a continuous inflow and outflow,** a building up and breaking down of components, never being, so long as it is alive, in a state of chemical and thermodynamic equilibrium but maintained in a so-called steady state which is distinct from the latter." (Ibid., p.39)

"**A system can be defined as a complex of interacting elements.** Interaction means that elements, p, stand in relations, R, so that the behavior of an element p in R is different from its behavior in another relation, R'. If the behaviors in R and R' are not different, there is no interaction, and the elements behave independently with respect to the relations R and R'. A system can be defined mathematically in various ways. For illustration, we choose a system of simultaneous differential equations." (Ibid.,p.45)

"**The organism is not a closed, but an open system. We term a system 'closed' if no material enters or leaves it; it is called "open" if there is import and export of material.**" (Ibid., p.121)

"Any system as an entity which can be investigated in its own right must have boundaries, either spatial or dynamic. Strictly speaking, spatial boundaries exist only in naive observation, and all boundaries are ultimately dynamic." (Ibid.,p.215)

DEDUCTIVE SYSTEMS THEORY

"The history of systems science has its beginnings in the years around 1950 (Hammond, 2003) with the work of founding fathers such as Bertalanffy, Wiener, Rapoport, Boulding and Miller. The emergence of systems thinking is closely linked with the endeavour to overcome previous boundaries within academia and practice (interdisciplinarity and transdisciplinarity)...The goal is to better understand how different sorts of systems work and how to deal with complex situations and reduce unwanted side effects...The so called **systems approach is often portrayed as a counter-current to the increasing fractionation of science into highly spezialized branches resulting in a breakdown of communication between the specialists.** (Rapoport, 1986, preface)"
('Understanding Systems Science: A Visual and Integrative Approach'. By Andreas Hieronymi.First published: 18 October 2013 https://doi.org/10.1002/sres.2215)

To achieve this goal, deduction is needed. More than 30 years ago, I applied the deductive method to teaching and published a paper, **'ON the Duality of Open-Close of Systems'** in *Studies in Philosophy of Science*, Jan, 1997. Years later, I put it into *Imagination Science.* (2010)

Bertalanffy characterizes GST (General System Theory) as 'a logicomathematical discipline, in itself purely formal but applicable to the various empirical sciences', use mathematical methods to explain, I put it in the category of reductionism.

"Cybernetics, e.g., proved its Impact not only in technology but in basic sciences, yielding models for concrete phenomena and bringing teleological phenomena-previously tabooed-into the range of scientifically legitimate problems; but it did not yield an all-embracing explanation or grand world view, being an extension rather than a replacement of the mecanistic view and machine theory (J. Bronowski, 1964)."

This is a passage from Bertalanffy's book (p.23), which seems is equally appropriate to comment on GST. Only when the system theory is based on holism can it stay away from the machine theory, and my theory explicitly rejects the popular concept of open & closed systems.

ARGUMENT

All systems are the products of the environment, and interact with the environment. There is no isolated or closed system without any connection to, or any exchange of matter & energy with environment, and the open system, as its opposite, cannot exist alone. In essence, the environment is a higher level system, and systems are hierarchical. All systems must exchange matter & energy with the external environment, which is the openness of them; All systems have boundaries (shells) and control mechanisms that control their inputs & outputs, which is the closure of the system. **All systems have the duality of openness and closure. The terms open and closed systems can only be relative, not absolute.**
Unlike Bertalanffy, **I think most systems are relatively closed.** For example, animals, which is regarded as a typical open system by Bertalanffy, need air, water as well as food all the time, and are open. But they're also closed in terms of selecting and controlling inputs. Atoms, molecules, planets, stars, and galactic vortexes are all intrinsically closed, that is, all of them have shells; their openness is secondary. **For all systems, excessive openness means**

decaying and death.

In reality, there is no isolated open & closed system, and the proposition that 'a closed system tends to maximum entropy' is not universal. A black hole are considered by ST to be a highly closed system from which not even light can escape. While CFT sees it as the fetus of a new celestial body, full of vigor and apparently devoid of entropy.

EMERGENCE.

"When electrons or atoms or individuals or societies interact with one another or their environment, **the collective behavior of the whole is different from that of its parts. We call this resulting behavior emergent.** Emergence thus refers to collective phenomena or behaviors in complex adaptive systems that are not present in their individual parts.

Examples of emergent behavior are everywhere around us, from birds flocking, fireflies synchronizing, ants colonizing, fish schooling, individuals self-organizing into neighborhoods in cities — all with no leaders or central control— to the Big Bang, the formation of galaxies and stars and planets, the evolution of life on earth from its origins until now, the folding of proteins, the assembly of cells, the crystallization of atoms in a liquid, the superconductivity of electrons in some metals, the changing global climate, or the development of consciousness in an infant.

Indeed, we live in an emergent universe in which it is difficult, if not impossible, to identify any existing interesting scientific problem or study any social or economic behavior that is not emergent." ('Emergence: A unifying theme for 21st century science'. By David Pines, Co-Founder in Residence, Santa Fe Institute. Santa Fe Institute, Nov 1, 2014. https://medium.com/)

"Emergence indicates that a system as a whole might have qualities or dynamics that its components on their own miss. A widely discussed example of this somehow surprising property is water. Water consists of oxygen and hydrogen, both of which are highly inflammable. Being put together however, these components interact and create water which extinguishes fire. Another simple example of an emergent system with eigenbehavior is the traffic jam, consisting of a multitude of individuals whose aim it is to get to their destination as fast as possible. In interaction however, they might cause the opposite." ('Closed & Open Systems: Definition & Examples'. By David Wood, Instructor. Study.com)

Methinks the most convincing example is **the vortex system.** The nebula is scattered, without light, and is caught in the vortex, becoming a red giant, a black hole and a blue giant in turn, which is shining brightly. Then, planets, satellites and their minerals and organisms are produced in turn, **which is the most common and perfect systematic evolution and the richest emergent activities in the universe.**

Different from D.Pines, I think all emergent behaviors of systems have leader and force. Gaseous oxygen and hydrogen form a liquid water system is led by oxygen, and the swirling force of atoms between them. The leader of 'the formation of galaxies and stars and planets' is light, it itself is cosmic force.

Without leader and force, there is no core and organization. They are just random piles of stuffs, like a heap of sands, a pile of garbage, a book of nonsense, etc., what I call disordered and lifeless loose systems, mostly artificial.

Fortunately, the universe is dominated by countless tightly knit systems of order, organization, vitality and self-perfection. The universe itself is such a system,of which man is the highest product.

The First Step of Creation: Choosing a Direction

"in geometrical diagrams, which have often a slight and invisible flaw in the first part of the process, and are consistently mistaken in the long deductions which follow. And this is the reason why every man should expend his chief thought and attention on the consideration of his first principles:- are they or are they not rightly laid down? and when he has duly sifted them, all the rest will follow." (Plato, Cratylus, 436d)

Creation is a keyword in this book. If readers arrive here, they may understand the importance of **choosing a direction, or, as the Greek thinkers put it, the first principal.** The previous section showed that a new kind of cosmology has come into being simply by changing the direction from infinite outward divergence to absolute centripetal contraction. The invention and innovation triggered by this can be said to be endless, and only some of these effects are listed in the previous section.

You may say that CFT has not yet been proven (Do conspicuous vortexes all over the sky have to prove their existence?), that this new theory is not necessarily good and does not rule out a pit. **If you have such doubts, you may as well wait and watch patiently, or participate in criticism.** Hawking mentioned three different arrows of time in chapter nine of *A Brief History of Time,* and they all point in the same direction. The third of these arrows is the cosmological arrow of time. Hawking clarifies "the direction of time in which the universe is expanding rather than contracting." (2016, p.164) He claims that the universe will shrink one day, but "it will not begin to shrink until at least ten billion years later," (p.169) so it is safe to follow the direction of standard cosmology.

This book also provides a new way to explore truth, namely, **obtaining truth through speculation and logical**

reasoning instead of observation and experiment. This is a way that most scientists shrug off, but it is indispensable to innovation.

In fact, this method is not new, it is a common method used by ancient Greek philosophers. Leucippus' vortex theory, Democritus' atomism and Euclidean geometry were created in this way. An outstanding example from the contemporary world is the twentieth-century American physicist Nikola Tesla, whose inventions are numerous, and many of which are based on speculation and logical reasoning, some of which are very precise. Einstein's theory of relativity was first and foremost a product of imagination and logical reasoning rather than observation and experiment. In addition, the idea of gravity, wormholes, black holes, singularity, neutron stars, antimatter, dark matter, and dark energy, along with string theory, inflation theory, the Big Bang theory, and so on, were all created in this way.

We may believe that logic is a powerful tool of wisdom, not limited to Aristotelian syllogism, and that **no theory can escape the scrutiny of logic.** "Reason is mistress and queen of all things." (Cicero) "that Logic which teaches how to use one's reason correctly in order to discover the truths of which one is ignorant;" (René Descartes) "Logic is the armory of reason." (Thomas Fuller) "Logic helps us to strip off the outward disguise of things, and to behold and judge of them in their own nature." (Isaac Watts, 17th-18th Century English minister) A logical theory cannot be easily overturned even if it lacks factual support and popularity. The illogical theory, however, popular at present, will collapse sooner or later.

Wisdom entails discrimination and the ability to choose, and these abilities rely on speculation and logical reasoning. **But logical reasoning is only a tool, above it, more important is the purpose and direction — the acme of wisdom.** In everything we do, the direction we choose is the most important thing. With the right direction, even for a slow walker, there is always the day when one will reach the destination. And for those who choose the wrong direction, all of their efforts will be in vain. The more effort that is put forth, the farther away such people will get from the destination. The best case is that one will achieve some isolated results, such as the discovery of something new, or put forward some specific methods such as the establishment of a practical mathematical model. However, it is not helpful to the realization of the goal, and even has the opposite effect. The typical type is the formula of universal gravitation — **without understanding the nature of gravity, it is impossible for anyone to have a correct gravitational model and formula.** It is as unreliable as a model and a formula that someone built by observing a UFO or a spook.   Quantum theory is now on the same path.

There is also a difference between high and low ends. **Wisdom entails every seeker of truth to know the highest end and pursue it, or at least not do the opposite.**

These are not empty phrase. An important achievement of human civilization for thousands of years has been to recognize the mission of human beings and the direction (end or goal) of the world.

Cosmic Evolution and Human Mission

I strike the stars with by sublime head. Horace

Without the ideal, the inexhaustible source of all progress, what would man be?
– Madame Delphine Gay de Girardin (GIGA Quotes)

All the products created by the universe, from atoms, to stars, to human beings, and from their structures, to movements, to shapes, are perfect and in perfect order. The universe has been running on a forward and upward track the whole time. This is the absolute law of the evolution of the universe, the law of 'perfection or death', and the highest manifestation of the cosmic spirit. Evils and atrocities of the world are manifestations of violating the laws of the cosmic evolution, that happen a lot in human society.

The earliest prophet of this was Longinus, a Greek who lived in Rome in the first century and taught eloquence. In his book *On the Sublime* (https://www.gutenberg.org/files/17957/17957-h/17957-h.htm), he pointed out that being vulgar and humble creatures is not the goal set by Nature for human beings. **Nature has birthed us in this universe, like putting us in a great arena, and wants us to be not only the audience of its great achievements but also its ambitious competitors.** Although the authenticity of this book and its author are controversial, later the British thinkers Francis Bacon and Sir James H. Jeans, the Prussian philosopher Immanuel Kant, the Italian thinker Giovanni Pico della Mirandola, the American physicist John Archibald Wheeler, and some authors, expressed similar ideologies by putting forth the concept of utopia, they believed that it was possible for man to reach the ideal, even the topmost state in the universe. In this regard, people should be highly confident; **"nothing is too high for the**

daring of mortals: we would storm heaven itself in our folly." (Horace) "By steps we may ascend to God." (John Milton, GIGA Quotes)

Humankind is the most advanced and perfect product of the evolution of the universe. The rest of the universe operates only in motion, change, and creation, not speech. Man naturally becomes the speaker and promoter of the universe.

The mission of the universe has thus become the mission of men. Man's task is to continue this eternal evolution of the universe through continuous creative activities. "Oh for a muse of fire that would ascend the highest heaven of invention!" (William Shakespeare) Following the goddess, man would also ascend this realm.

This is an unprecedented cause of creation, so a corresponding philosophy with creation as its aim needs to be established. This means a philosophical revolution, because previous philosophies, including Plato's, lacked the idea of creation. Almost all religions believe that creation is God's task, but there are also many people in the world who believe that human beings have creativity, and lots of things have been created that are not found in nature. Even religions and Gods are human beings' creative works. In religion, there are a few devout believers who believe that it is not enough for people to worship God alone, creating brilliant and magnificent things, showing miraculous signs, and completing the unfinished jobs of God in the world shows the greatest reverence for God.

This task has actually started, the philosophy of cosmic force, which was established before the establishment of the cosmic force cosmology. It is a unique philosophy that holds high the banner of creation. It claims that the purpose of human beings is to create, and that the purpose of philosophy is also to create. It requires people to recognize the best characters, ideas, and works of art in the world, and strive to transcend them. **Creation is the source of all beautiful things and is the eternal motivating force of an ever-advancing world.** Cosmic force theory has established its unique position in philosophy by way of its distinctive concept of creation.

The last words I wrote in *The Plato Code* more than ten years ago are fitting here:

Today, humans can proudly say to God:

"O Lord, thou hast created man and all things which are changing and perishable, but **we humans have created many immortal and eternal things!**"

Why Do I want to Publish This Book?

More than two hundred years ago, two things awed Immanuel Kant: the starry sky above him and the moral law within him.

Today, there are two things that shock me most:
One. The vortexes that are ubiquitous in the sky are generally ignored. Most people would rather believe that the universe is expanding than that it is shrinking everywhere under the force of the vortex.
Two. One of the grand cosmic forces — nuclear force — was discovered not for the purpose of construction but for self-destruction. Thousands of social elites are working day and night to study and upgrade nuclear weapons so that they can one day be used to destroy the world and annihilate humankind in the fastest, most effective, and most thorough way.

Sane people know that this is the most absurd and bloodiest behaviour. We must try our best to stop it.
When we are facing ignorance and evil, the usual method is tit-for-tat, or legal sanctions. I have more appreciation for the ancient Greek way, which is dedicated to artistic activity. **Art is the creation of beauty & goodness, including philosophy, science, and the creation of all good things.** The Greek's conquered the world in this way. Even if foreign invaders could occupy Greek territory, they eventually surrendered to the seat of the Greek art goddess. The truth is that beauty & goodness — Greek: kalos kai agathos — is the inner aspiration of most people; it is the highest universal value and the world's ultimate goal, so its force is extremely powerful and invincible.
Ignorant people think that Greek civilization is dead and no longer sheds any light on today world. They fail to see

that the Greek's are the authors of the philosophy of beauty & goodness, which is immortal. **In fact, the world hasn't really known the power of Greek art until now.**

Law can restrain a man's conduct, but it cannot improve his character. War can quickly change people's thinking and behavior, but it often acts on the current generation, and its destructive effect is so great that we must try to prevent it. History tells us that **artistic creation is the broad avenue for human ascension and world progress.**

Art is not only fine music, painting, sculpture and dance, but also includes all fine thoughts, speeches, institutions, deeds. If there are many people in the world who love art, devote themselves to creative activities and constantly provide beautiful and good works, then the world will naturally move towards harmony and friendship without strict laws and tough management rules.

To support artistic creation, I repeat the key ideas of this book here:

- The first law of cosmic evolution: **aggregation gives rise to all things, while separation destroys all things.** This determines that the universe can't be expanding continuously.

- Energy is the cause of force, the force of all forces, the inner essence of everything; Force is the application of energy and phenomenon. Without energy, without force. The universal form of energy is vibration & resonance, and the ultimate source of all energies and forces is **Cosmic Energy, LIGHT.**

- Please enjoy the cosmic spirit brand products: darkened light theory, light ball theory, light field theory, shell theory, cosmic force vortex theory and cosmic force theory.

- No object can travel faster than the speed of light, but **both light balls and the light field are expanding at twice the speed of light at all times.** It is the most brilliant wonder of our cosmos.

- An incredible fact. The glowing filament of the lightbulb, the sunshine, and the SHC have one thing in common. **They all are the result of the electrons leaving their atoms and the object becoming a plasma.**

- Free particles glow when they are compressed into plasma. Therefore, we can boldly claim that stars glow without thermonuclear reaction or nuclear fusion, they won't age from 'hydrogen burning'. **The main stuff of plasma is hydrogen ions, which is self-luminous.** Incandescent lamps only glow, violent resonance of plasma, not burn. This is exactly the case with the sun.

- If you believe that temperature is nothing more than an indicator of the intensity of the vibrations of ions, atoms and molecules, in direct proportion to their speed of motion and in inverse proportion to their density. You won't be surprised to learn that **the sun's core is cooler than its surface,** and you'll be well aware of the nature of sunspots.

- Why do stars move in circles instead of straight path? Shells do it. **Shells force all linear motion into curvilinear motion, centrifugal motion into centripetal motion, and divergent rotational motion into concentrated convergent vortex motion.** With the help of advanced means, many scientists have observed the vortex inside atoms, which is undoubtedly the act of shells. **Vortex is an infinite tightening force**, and it is much better to explain the solid polymerization of nucleons than to explain it in terms of vague strong force & gluons.

- Newton defined gravity as the centripetal force, "that by which bodies are drawn or impelled, or any way tend, towards a point as to a centre." (The Principia 3) This is an extremely laborious task, and in fact no any force can do it. **Shells allow all celestial bodies to move in stable and precise orbits naturally, without extra energy at all.**

- The space is vast, all vortexes evolve in their own shells, and communication between them is rare. Besides a few meteors and debris, **most of the objects on stars and planets are native products of their own,** not 'import goods'. Their earth, water, air, metals, and living things are all transform from their own plasma.

- Shock: astronomers are always searching everywhere for facts but turn blind eyes and deaf ears to a downright fact: **Heavenly vortexes** all over the sky!

- Gentle reminder.
  The indispensable condition of gravity is action at a distance, which Newton dismissed as 'great an absurdity'. **Our textbooks today still treat gravity as a cosmic truth, neglecting action at a distance that fills everywhere in the curved space.** "Newton's followers, of course, did not fail to notice that his system had emptied out space. Having fewer scruples, they became more Newtonian than Newton." (Wilczek, 2008, p.78)

- Why does Newton's law work so well while this book denies gravity?
  Look at two kinds of gyroscopes: electric gyro and

whipping gyro. We can design them to look the same, but their principles are completely opposite. The former, with its power at the center and energy flowing inside out, is a centrifugal and divergent motion that fits standard cosmology. The power of the latter is external, and the energy flows outside in. It is a centripetal convergent motion, which is suitable for CFC. If you ignore that all and just calculate their velocities, the two have no significant difference. That's the Newtonian approach to experimental philosophy. **Newton actually treated vortex forces as gravitational forces. The spins and orbits of the stars are the most common and steady vortex motions.** Gravity can't make the gyro or planet rotate, it will only force them to stop rotating.

- Marvelous Speculative Philosophy➤The Hellenic Thinker's Eight Hypotheses
  1. Aristotle's Ether
  2. Heraclitus' Cosmic Fire
  3. Leucippus' Vortex
  4. Democritus' Atomism
  5. Euclidean Geometry
  6. Empedocles' Evolution
  7. Aristarchus' Heliocentrism
  8. Archimedes' Buoyancy

  They construct these hypotheses in primitive conditions, relying on speculation and deduction. The last five have been tested validated and applied by the world, while the first three have been debating and discussing until the publication of this book. I unite the three with LIGHT in this book, thus they are validated, fully empowered, and become **the coping-stones of CFC**.

- Why are there lots of cumbersome quotations in this book?

> **"We ought to read the writings of the ancients,** for it is of great advantage to be able to make use of the labours of so many men. We should do so both in order to learn what truths have already been discovered and also to be informed about the points which remain to be worked out in the various disciplines." (Descartes, 1985, p.13)

This instance also shows that **CFT is a dialogue transcripts, the work of numerous intelligent minds in the world. I'm just the lead author.**

A motto. The greatest minds of the world have long left us the wisdom for solving the most abstruse cryptograms in various forms. Search and we will find.

OUTLOOK

At the end of this book, I like to introduce a book I wrote 10 years ago, *From Fishing Ape to Genius* (2013), which is a unique theory of evolution.

In my opinion, evolution is definitely humanism, which is about the origin and evolution of human consciousness, whose core factor is sexual love. Like other animals, early fishing apes understood only sex, not love. **Love was gained in their later aquatic life and in face to face intercourse.** Love causes consciousness, emotion and language to occur, from low to high, from individual to society, which is a process of continuous expansion and reinforcement. At the beginning of civilization, human consciousness break through the boundaries of individual & family, and today it extends further to the globe and the cosmos.

Man is a product of the cosmos. Unlike other products, **man has spiritual self-consciousness, the ultimate product of cosmic spirit. He can even transform intangible things into tangible and infinite things into finite. Men rely on their consciousness to explore the cosmos, recreate various cosmos, name all things, and**

measure things values and status. As a result, man has become the soul of the cosmos.

On this road, I am a latecomer, and there are many pioneers before. Among them, Francis Bacon is the most outstanding one. He encouraged people to explore the cosmos and correctly identified LIGHT as the bullseye.

"Further, it will not be amiss to distinguish the three kinds and, as it were, grades of ambition in mankind. The first is of those who desire to extend their own power in their native country, a vulgar and degenerate kind. The second is of those who labor to extend the power and dominion of their country among men. This certainly has more dignity, though not less covetousness. But **if a man endeavor to establish and extend the power and dominion of the human race itself over the universe, his ambition (if ambition it can be called) is without doubt both a more wholesome and a more noble thing than the other two.** Now the empire of man over things depends wholly on the arts and sciences. For we cannot command nature except by obeying her.

Again, if men have thought so much of some one particular discovery as to regard him as more than man who has been able by some benefit to make the whole human race his debtor, how much higher a thing to discover that by means of which all things else shall be discovered with ease! And yet (to speak the whole truth), as the uses of **light** are infinite in enabling us to walk, to ply our arts, to read, to recognize one another — and **nevertheless the very beholding of the light is itself a more excellent and a fairer thing than all the uses of it** — so assuredly the very contemplation of things as they are, without superstition or imposture, error or confusion, is in itself more worthy than all the fruit of inventions." (Bacon, 2018, BOOK I,129)

This book can be seen as a substantial step towards the 'third ambition' proposed by Bacon 400 years ago, a concrete program of practice. The highest secret of the cosmos is revealed here. I expect a large number of participants in the future.

Appendix: Firmly Support Absolute Justice ACT (AJA), Strongly Curb Nuclear War!

The weapon of criticism certainly cannot replace the criticism of weapons, the material forces can only be destroyed by material forces.
Karl Marx: *An Introduction of Critics of Hegel's Philosophy of Right*

I am an idealist. I believe that the power of beauty and goodness will prevail on Earth and that there will be an ideal age in the world.

A major scientific achievement in the last century was the development and utilization of nuclear energy. It is the cosmic force, the gift of nature and a precious public property of mankind. Borrowing Immanue Kant's 'Absolute Order' (Deutsch, kategorisches imperative), a new absolute order is that **cosmic force can only be used for the happiness of mankind and cannot be allowed as a power to destroy the world.**

Hold to this order, nuclear weapons were made, and used to end World War II. It was an **AJA (Absolute Justice Act)** and was widely approved by people around the world.

Similarly, in accordance with this Order, continuing to produce and upgrade nuclear weapons after the WWII, whether intended for attack or for defence, is no longer an act of justice, but an act of cowardice and evil, which threatens the very existence of everyone and has become **public enemy Number One of the world.** To eliminate nuclear weapons and stop nuclear war is the right and duty of everyone, mine included.

Eliminating nuclear weapons and stopping nuclear war seems to be a formidable task. With AJA, however, the task becomes simple and realistic.

Here, I hope you will not be offended by the idea of the **Absolute,** as if I were imposing an abstract and impractical concept on you. Hegel said, "Common fancy puts the Absolute far away in a world beyond. The Absolute is rather directly before us, so present that so long as we think, we must, though without express consciousness of it, always carry it with us and always use it." (*Shorter Logic*, §24)

Both the Absolute and the Necessity mean a force which must happen in and for itself, which is what the Greeks called destiny, the materialists called a motion independent of man's will, the theologians called divine will. Anyone can go his own way, ignore it, or deliberately fight it, the result is bound to be miserable. If he win, even worse.

AJA is such a movement, and it will go on and do it itself without my article. My purpose in writing this article is to make it happen smoothly. It is the task of man, not God, and depends on man to complete it. If more people become aware of this movement and join it, the fate of MAD (mutually assured destruction) can be avoided.

The goal of AJA is, to completely destroy all nuclear weapons, crush nuclear bomb factories and their launching facilities and make all of them disappear from the Earth forever. Correspondingly, there are two means to this goal, spiritual and material.

I believe that spiritual weapons are ten thousand times more powerful than nuclear bombs. The power of theory and public opinion is stronger than that of mass demonstrations and protests. It is essential to make people realize that developing nuclear weapons is an evil and dirty act, to ask scientists to refuse to be involved in this kind of activity, to ask politicians to sign the nuclear ban treaty and do it right away, and to make people hate nuclear war from the bottom of their hearts — this is the most fundamental goal.

Many heads of state and political leaders have been trying to defend human rights, even to the point of believing that human rights outweigh sovereignty and denouncing all violations of human rights. Nuclear weapons are the most powerful weapon against human rights, and if they can put their claims into practice, nuclear weapons will soon disappear from this planet.

Philosophy and scientific theory are the most powerful mental weapons, and if they work, nuclear war could be eliminated in this gentle way.

However, history and experience tell us not to be too optimistic here. There are too many nuclear bomb enthusiasts in the world at present. They will not give up their bombs easily. Even if the United Nations ratifies the Treaty on the Prohibition of Nuclear Weapons and declares it enters into force on January 22, 2021, they do not necessarily have to abide by it. When righteous people are trying to achieve their purpose, as Karl Marx said, spiritual force is not enough, material force will have to be used as well — it's actually already being used.

The year 2020 marked a milestone in the anti-nuclear war movement when Iran's nuclear facilities were destroyed in mysterious explosions and the top nuclear scientist of the country was assassinated, bringing an end to the country's nuclear program.

After this, the Iranian government vowed revenge, then backed down. I think there are two considerations that keep them from taking revenge.

First, the development of nuclear bombs is an evil act that God and man generally oppose, especially in this sensitive area of the Middle East. It is AJA to destroy its nuclear base and kill the nuclear expert.

Second, the benefits for Iran. AJA passes unimpeded in the world; it is an absolutely reprehensible act, good for all, including the Iranian people at large. The development of nuclear weapons is a heavy burden for all countries and does nothing to help their security. If the Iranian government gives up its nuclear program from then on, this will reduce its international pressure, improve relations with major powers, boost its economy and make the country more secure. I believe most Iranians are also opposed to developing nuclear weapons.

The biggest implication of the Iran incident is that it

declares that the nuclear crisis is over and that the threat of nuclear weapons to the world has been ended. This incident shows that only out of survival instinct and self-defence, will people take action to destroy nuclear facilities and eliminate nuclear experts at a critical juncture, instead of waiting for their opponents to elaborate terrorist weapons and destroy themselves.

Although nuclear weapons are extremely destructive, they are inseparable from bases and launching tools. I foresee with optimism that as long as international tensions and relations deteriorate, conventional wars to attack each other's nuclear bases, depots and launch tools will begin prematurely, and the scariest nuclear bombs will explode in situ before they can be launched. Even if a country could successfully launch a nuclear bomb, it would not be able to cause a major disaster because it would face international sanctions and that day would be its doomsday.

Legal means, plus AJA, are the surest way to prevent nuclear war. **People do not have to beg for mercy from the rulers of nuclear powers. They only need to support and join all countries, groups, organizations and individuals in sabotages of nuclear weapons and facilities, and make rulers and experts who are addicted to developing nuclear bombs feel threatened day and night, so that nuclear war can be avoided.**

In my new book, *An Independent Thinker Growing in Shells: An Autobiography,* published this year, I also propose that the United States should take the lead in this effort, which is its duty, as the US also has enough strength to fulfil the duty. This is the better way.

References

Aristotle. (2021). *Metaphysics*, translated by W. D. Ross. Global Grey ebooks.

Aristotle. *Physics*, translated by R. P. Hardie and R. K. Gaye. Provided by The Internet Classics Archive.

Asimov, I. (1972). *Asimov's Guide to Science*. BASIC BOOKS, INC.

Asimov, I. (1993). *Asimov's New Guide to Science*. Penguin.

Asimov, I. (1976). *Please Explain* (Chinese version). Dell Publishing Co., Inc.

Bacon, F. (2018). *THE NEW ORGANON OR TRUE DIRECTIONS CONCERNING THE INTERPRETATION OF NATURE*. Global Grey ebooks.

Ball, P. (2018). *Through Two Doors at Once: The Elegant Experiment That Captures the Enigma of Our Quantum Reality.* Anil Ananthaswamy Dutton. Nature 560, 165.

Bally, J & Reipurth, B. (2006). *The Birth of Stars and Planets*. London, UK: Cambridge University Press.

Barrow, J. D. (2012). *The Book of Universes*. New York, NY: Vintage.

Beaty, William J. (1995). Misconceptions Spread by K-6 Textbooks: 'Electricity'.
http://amasci.com/miscon/eleca.html

Bertalanffy, L. von. (1969) *GENERAL SYSTEM THEORY. Foundations, Development, Applications.* University of Alberta, Edmonton, Canada. GEORGE BRAZILLER, New York.

Brooks, R. A. (2016). *Fields of Color: The Theory That Escaped Einstein*. Silver Spring, MD: Universal Printing, LLC.

Clegg, B. (2018). *Gravitational Waves: How Einstein's Spacetime Ripples Reveal the Secrets of the Universe*. London, UK: Icon Books.

Davies, P. C. W. (1986). *The Ghost in the Atom*. London, UK: Cambridge University Press.

Descartes R. (1985). *The Philosophical Writings of Descartes*, Volume 1. Translated by J. Cottingham, R. Stoothoff, D. Murdoch.© Cambridge University Press.

Descartes R. (2004). *The World and Other Writings*. Edited by S.Gaukroger. © Cambridge University Press.

Doyle, A. C. (2008). *A Study In Scarlet*. Gutenberg EBook

Durant W. (1991). *The Story Of Philosophy: The Lives and Opinions of the Greatest Philosophers.* TIME INC. NEW YORK.

Einstein, A. (1999). *Autobiographical Notes*. Open Court.
Einstein, A. (2005). *The New Quotable Einstein*. Princeton University Press.
Einstein, A. (2006). *Relativity* (Chinese version). Chongqing, China: Chongqing Publishing House,CHN.

Engels, F. (1883). *Dialectics of Nature*. Transcribed 1998/2001 for MEIA by slr@marx.org,jjazz@hwcn.org. http://www.marxists.org/archive/marx/works/1883/don/index.htm

Ferris, T. (2003). *Coming of Age in the Milky Way*. New York, NY: Harper Perennial.

Ford, K. W. (2005). *The Quantum World: Quantum Physics for Everyone*. Cambridge, MA: Harvard University Press.

Gamow, G. (1947). *One, Two, Three… Infinity: Facts & Speculations of Science*. New York, NY: Viking.

Halpern, P. (2004). *The Great Beyond*. Hoboken, NJ: Wiley & Sons.

Hawking, S & Mlodinow, L. (2011). *The Grand Design*. London, UK: Bantam.

Hawking, S. (2016). *A Brief History of Time*. London, UK: Bantam.

Hegel, G.W.F. (2005). *Shorter Logic* (Chinese version). Beijing, China: The Commercial Press.

Heraclitus. (2016). *Fragments*. Translated by John Burnet, Arthur Fairbanks, and Kathleen Freeman. antilogicalism.files.wordpress.com

Heisenberg, W. (1971). *Physics and Philosophy: The Revolution in Modern Science*. Crows Nest, Australia: Allen & Unwin.

Horgan, J. (2015). *The End of Science: Facing the Limits of Knowledge in the Twilight of the Scientific Age*. La Vergne, TN: Ingram.

Kaku, M. (2012). *Physics of the Future: How Science Will Shape Human Destiny and Our Daily Lives*. Random House.

Kant, I. (1972). *Universal Natural History and Theory of the Heavens* (Chinese version). Shanghai, China: Shanghai People's Publishing House, CHN.

Kippenhahn, R. (1993). *100 Billion Suns: The Birth, Life, and Death of the Stars*. Princeton, NJ: Princeton University Press.

Kline, M. (1983). *Mathematics: The Loss of Certainty*. London, UK: Oxford University Press.
Kline, M. (1953). *MATHEMATICS IN WESTERN CULTURE*. UK: Oxford University Press.

Kragh, H. (2002). *Quantum Generations: A History of Physics in the Twentieth Century*. Princeton, NJ: Princeton University Press.

Kuphaldt, Tony R. (2021). *Lessons In Electric Circuits*. http://www.ibiblio.org/kuphaldt/electricCircuits/

Laërtius, D. (2006). *The Lives and Opinions of Eminent Philosophers*. Whitefish, MT: Kessinger Publishing Co.

Lipton, B.H. (2008). *The Biology of Belief: Unleashing the Power of Consciousness, Matter and Miracles*, Hay House UK.

Melia, F. (2003). *The Edge of Infinity: Supermassive Black Holes in the Universe*. London, UK: Cambridge University Press.

Newton, I & Thayer, H. S. (2001). *Newton's Philosophy of Nature: Selections from His Writings* (Chinese version). Shanghai, China: Shanghai Translation Publishing House, CHN.

Newton, I. (2020). *The Mathematical Principles of Natural Philosophy (The Principia)*, translated by Andrew Motte, Global Grey ebooks.

Planck, M. K. E. L. (1959). *The New Science*. Meridian

Books.

Plato. (n.d). *Laws*. (B. Jowett, Trans)

Plato. (n.d). *The Republic*. (B. Jowett, Trans)

Plato. (n.d). *Timaeus*. (B. Jowett, Trans)

Reichenbach, H. (1951). *The Rise of Scientific Philosophy*. California, CA: University of California Press.

Rosenblum, B & Kuttner, F. (2011). *Quantum Enigma: Physics Encounters Consciousness*. Los Angeles, CA: Prelude Press.

Smolin, L. (2007). *The Trouble with Physics*. Boston, MA: Mariner Books.

Solatle Lu. (2008). *The Plato Code*. (Chinese version) Yunnan People's Publishing House.

Thorne, K. S. (1994). *Black Holes and Time Warps: Einstein's Outrageous Legacy*. London, UK: Pan Macmillan.

Tyson, N. deGrasse & Goldsmith, D. (2006). *Origins: Fourteen Billion Years of Cosmic Evolution*. London, UK: W. W. Norton & Co.

Watson, A. (2004). *The Quantum Quark*. London, UK: Cambridge University Press.

Weinberg, S. (1993). *Dreams of a Final Theory: The Search for the Fundamental Laws of Nature*. New York, NY: Vintage.

Wilczek, F. (2008). *The Lightness of Being: Mass, Ether, and the Unification of Forces*. New York, NY: Basic Books.

【中文书名】**原力宇宙学**——21世纪科学-哲学革命宣言

www.ingramcontent.com/pod-product-compliance
Lightning Source LLC
Chambersburg PA
CBHW031602210526
45464CB00004B/1401